Selected titles by Norman F. Cantor

The Civilization of the Middle Ages

Inventing the Middle Ages

Encyclopedia of the Middle Ages

Imagining the Law

Medieval Lives

The Medieval Reader

The Sacred Chain: The History of the Jews

Dec.
1350

June
1350

Dec.
1349

June
1349

Dec.
1348

June
1348

Dec.
1347

THE SPREAD OF THE BLACK DEATH ACROSS
EUROPE IN THE FOURTEENTH CENTURY.

Graham Twigg, The Black Death, *1984*

In the wake of the plague

THE BLACK DEATH AND THE WORLD IT MADE

◆ NORMAN F. CANTOR ◆

POCKET BOOKS

LONDON · SYDNEY   NEW YORK · TOKYO   SINGAPORE   TORONTO

First published in the USA by Simon & Schuster Inc, 2001
First published in Great Britain by Simon & Schuster UK Ltd, 2001
This edition first published by Pocket Books, 2002
An imprint of Simon & Schuster UK Ltd
A Viacom Company

5 7 9 10 8 6 4

Simon & Schuster UK Ltd
Africa House
64-78 Kingsway
London WC2B 6AH

www.simonsays.co.uk

Simon & Schuster Australia
Sydney

A CIP catalogue record for this book is available from
the British Library

ISBN 0-7434-3035-2

Printed and bound in Great Britain by
Cox and Wyman Ltd, Reading, Berkshire

Contents

To my family

PART I

Biomedical Context

〜〜

CHAPTER ONE

All Fall Down

IN THE SIXTH MONTH OF THE new millennium and new century, the American Medical Association held a conference on infectious diseases. Pronouncements by scientists and heads of medical organizations at the conference were scary in tone. Infectious disease was the leading cause of death worldwide and the third leading cause in the U.S.A., it was stressed. The situation could soon become much worse.

As the world becomes more of a global village, said one expert, infectious disease could by natural transmission become more threatening in the United States. Here monitoring is lax because of a mistaken belief that the threat of infectious disease has been almost wiped out by antibiotics.

Bioterrorism presented a further and much greater possibility of terrible outbreaks of pandemic in the United States. *The New York Times* reported: "A speaker at the meeting warned that the healthcare system in the United States was not prepared for a bioterrorist attack, in which hundreds or thousands of people might flood hospitals, needing treatment for diseases: anthrax, plague, or smallpox, which most doctors in this country have never seen."

In the same week as this AMA conference and its Cassandra-like speeches, the *NBC Nightly News* featured a brief segment showing American biochemists helping their Russian counterparts clean up and close down a large germ warfare factory. The TV correspondent remarked that the Russian plant had been capable of producing far more than the minimum required for effective biochemical warfare. He did not pursue the obvious questions of whether the Russians had been exporting the plants' surplus to Iraq, or if this was only one of several Russian germ warfare factories and whether the others may still be operating.

That *The New York Times* report was tucked away on page fifteen of its National Edition and that NBC News devoted all of four minutes to the Russian disease factory indicate that the problem of infectious disease and its pandemic threat to American wellbeing is still regarded as a marginal matter. By the time the next president of the United States finishes his term, it could be the most visible problem facing American society, similar to the biomedical crisis of late medieval Europe, England in particular.

In the England of 1500 children were singing a rhyme and playing a game called "Ring Around the Rosies." When I grew up in Canada in the 1940s children holding hands in a circle still moved around and sang:

> Ring around the rosies
> A pocketful of posies
> Ashes, ashes
> We all fall down

The origin of the rhyme is the flulike symptoms, skin discoloring, and mortality caused by bubonic plague. The children were reflecting society's efforts to repress memory of the Black Death of 1348–49 and its lesser aftershocks. Children's games were—or used to be—a reflection of adult anxieties and efforts to pacify feelings of fright and concern at some devastating event. So say the folklorists and psychiatrists.

The meaning of the rhyme is that life is unimaginably beautiful—and the reality can be unbearably horrible.

In the late fourteenth century a London cleric, who previously served in a rural parish and who is known to us as William Langland, made severe reference to the impact of infectious diseases "pocks" (smallpox) and "pestilence" (plague) in *Piers Plowman,* a long, disorganized, and occasionally eloquent spiritual epic. As translated by Siegfried Wenzel:

So Nature killed many through corruptions,
Death came driving after her and dashed all to dust,
Kings and knights, emperors and popes;
He left no man standing, whether learned or ignorant;
Whatever he hit stirred never afterwards.
Many a lovely lady and their lover-knights
Swooned and died in sorrow of Death's blows. . . .
For God is deaf nowadays and will not hear us,
And for our guilt he grinds good men to dust.

The playing children, arms joined in a circle and singing "Ring Around," and the gloomy, anguished London priest were each in their distinctive ways trying to come to psychological terms with an incomparable biomedical disaster that had struck England and most of Europe.

The Black Death of 1348–49 was the greatest biomedical disaster in European and possibly in world history. Its significance was immediately perceived by the wise Arab historian Ibn Khaldun, writing a few years later: "Civilization both in East and West was visited by a destructive plague which devastated nations and caused populations to vanish. It swallowed up many of the good things of civilization and wiped them out in the entire inhabited world." A contemporary Florentine writer referred to "the exterminating of humanity."

A third at least of Western Europe's population died in what contemporaries called "the pestilence" (the term the Black Death was not invented until after 1800). This meant that somewhere around twenty million people died of the pestilence from 1347 to 1350. The so-called Spanish influenza epidemic of 1918 killed possibly fifty million people worldwide. But the mortality rate in proportion to total population was obviously relatively small compared to the impact of the Black Death—between 30 percent and 50 percent of Europe's population.

The Black Death affected most parts of the Mediterranean world and Western Europe. Ingmar Bergman's 1957 film *The Seventh Seal* depicts the impact of the Black Death on Sweden. In Bergman's view the Black Death, which reached Sweden by 1350, caused an era of intense pessimism and widespread feelings of dread and futility.

But the great medical devastation hit no country harder than England in 1348–49 and because of the rich documentation surviving on fourteenth-century England it is in that country that we can best examine its personal and social impact in detail. Furthermore, there were at least three waves of the Black Death falling upon England over the century following 1350, nowhere near as severe as the cataclysm of the late 1340s, whose severity was unique in human history. But the succeeding outbreaks generated a high mortality nonetheless.

The population of England and Wales in the thirteenth

century had doubled. Unusually warm weather together with adequate moisture produced bumper crops and the generous food supply moderated the death rate. Then the downswing of the Malthusian cycle common to premodern rural societies set in.

Due to famines in the second decade of the fourteenth century the English population had begun to recede from its medieval peak of six million in 1300. But it was the Black Death that principally caused the demographic crash and the road back was slow and very long. When the English population began to rise significantly in the later seventeenth century there was yet another and final outbreak of the terrible pestilence in 1665 as graphically imagined by the journalist Daniel Defoe (author of *Robinson Crusoe*) in his *Journal of the Plague Year* (1722).

The level of English and Welsh population attained in 1300, close to six million people, was not reached again until the mid–eighteenth century.

Recently there have appeared in scientific journals and in the press articles and stories about diseases and pandemics in modern times that raise remarkable parallels with or connections to the Black Death and offer new perspectives on the fourteenth-century devastation. But there will likely always be a degree of uncertainty about the clinical history of the Black Death because of severe limitations of the fourteenth-century medical profession in diagnosing the ailments of its patients.

Fourteenth-century medicine was not without accomplishment. It could amputate limbs and normally cauterize the wounds in an effective manner. It had precious knowledge of herbal remedies for headache, minor stomachaches, menstrual cramps, and other marginal afflictions, possibly including psychological depression. But it was impotent in the face of an epidemic.

Medieval physicians still followed the theories of the second-century Greek doctor Galen, which attributed disease to imbalance in the bodily conditions, or "humours," of an individual. The main instrument of diagnosis was eyeballing the color and consistency of urine.

The prime remedies for illnesses involved restoration of putative bodily balance through purgation (enemas) or bloodletting. Drawing blood from a sick patient was considered a credible remedy until the nineteenth century. Cleaning the bowels was thought to have a curative effect. Enemas are still a popular home remedy. Nineteenth-century medicine introduced antiseptic surgery and anesthesia and smallpox inoculation but in the face of a pandemic outbreak was not much better off than the physicians of fourteenth-century England.

Faced with a worldwide outbreak of what was arbitrarily called Spanish influenza in 1918, which killed fifty million people within a year, the early twentieth-century medical profession was not much more effective in terms of diagnosis and cure than its medieval counterpart facing the Black

Death. Essentially the flu pandemic of 1918 came and went without anyone knowing why, in spite of the capacity to see under a microscope some viruses and bacteria that were totally invisible to the physicians of the fourteenth century. Recently, DNA analysis has begun on cell tissue taken from 1918 graves in Spitzbergen and Alaska.

After surveying what recent biomedical science tells us about the Black Death, this book studies the Black Death in two ways. It aims to show how the great biomedical devastation affected particular individuals, both victims and survivors, families, institutions, cultures, and social groups. It tries existentially to communicate the experience of this terrible ordeal, which may have some parallels in human society in coming decades.

This is a microcosmic closeup perspective on the Black Death. The second perspective is at the macrocosmic level. This book places the fourteenth century in context of the long history of such fearsome outbreaks of infectious disease, drawing upon our increasing knowledge of the history of medicine.

On the microcosmic level we will learn what happened to key individuals in a society overwhelmed by biomedical devastation. On the macrocosmic level, we will gain insight into the history of the human race from its beginning millions of years ago into the third Christian millennium.

CHAPTER TWO

Rodents and Cattle

❦❧

IN SPITE OF THE INCAPACITY OF THE medieval medical profession to describe securely the symptoms and course of the Black Death, historians of medicine and society have been able to determine that it involved at least bubonic plague, the same pandemic that had devastated the East Roman or Byzantine Empire in the sixth century A.D. and invaded the whole Mediterranean world in the third century or even earlier. The only big question on the medical side of the Black Death is whether bubonic plague was exclusively the cause of the devastation of the 1340s or whether another disease was simultaneously occurring in some parts of Europe, and particularly in England.

Bubonic plague is a bacillus carried by parasites on the backs of rodents, principally but not exclusively in the Middle Ages, the species of black rat. The black rats and the plague parasites residing on them could have been disseminated by shipping in international trade. The port of Bristol was the major initial point of entry for the pestilence into England.

It is this provocative picture of these rodents scurrying inland from port cities and making long journeys through the countryside at great speed so that most of Western Europe was in pandemic conditions within a year of initial contact that raises skepticism about the conventional account of the Black Death's exclusive identification with bubonic plague.

When a human contracts bubonic plague without antidote (not available until the applications of antibiotics in the 1940s), there is a four out of five probability that he will die within two weeks. The first stage is marked by flulike symptoms, normally accompanied by high fever. In the second stage, buboes—that is, black welts and bulges—appear in the groin or near the armpits. (Except about 10 percent of plague victims. In these unfortunate men and women the buboes develop intra-abdominally—that is, internally—and are only seen in autopsies.)

The buboes first grow as dark accretions on the skin. They vary in size from one to ten centimeters, but are all extremely ugly and extremely painful. Diarrhea and vomiting also accompany this, the crisis stage of the plague. Its incubation period, marked by fever, runs from two to eight days.

The third—and often fatal—stage of the plague is respiratory failure (pneumonia).

Today a patient is likely to recover if treated with antibiotics during the first two stages; if the disease reaches the third stage, antibiotics may not work.

Forty years ago historians believed that bubonic plague stopped affecting Europe in the eighteenth century because one species of rodent, the black rat, was replaced by another species, the gray rat. Even if this were true, which is not likely, it would not account for the disappearance of the plague, because the disease can be carried by any rodent and, today's scientists believe, by cats, of which there were plenty in the eighteenth century.

Moreover, there are peculiarities about the spread of the Black Death if it was exclusively bubonic plague that was involved. In 1984 the British zoologist Graham Twigg pointed out that the plague's impact, at least in England, was as severe in some thinly populated rural areas as in thickly settled areas. The pestilence produced almost as high a level of mortality in the winter months as in summer. These qualities do not easily conform to the view that the Black Death was exclusively bubonic plague: parasites on the backs of rats in thinly settled areas and severe impact in cold weather are not in keeping with the common activity of fleas.

Medical historians such as Twigg also noted that mortality tales of the period around 1350 frequently described a death that occurred within three or four days of incubation, much too rapid for the much longer three-phase course of the bubonic plague. Some patients died without fever and without the buboes or welts on groins or around armpits, and to explain their deaths it was proposed, in what is still a minor-ity opinion—although one rapidly gaining strength—that the Black Death involved or was even exclusively a rare viru-lent antihumanoid form of cattle disease, namely anthrax.

Both anthrax and bubonic plague begin with similar flu-like symptoms, and the two diseases could have been con-flated by contemporary doctors. And it is not hard to perceive how this anthrax-based plague—if Twigg's theory is correct—could have been spread. As Europeans cleared forests for more arable land in the thirteenth century, they did not attenuate their passion for red meat, even though the supply of wild game diminished with the forest clearing. There was an enormous increase in cattle ranching, raising of herds of beef cattle in congested conditions both on the great open ranges of northern England and the small pas-turages in the southern farmlands.

Before the widespread immunizing inoculation of cattle herds in the 1950s, infectious epidemics of anthrax murrain (cattle disease) were a constant threat in cattle ranches in the transatlantic world. Modern outbreaks of infectious disease among cattle, whether rinderpest in Rhodesia in the 1890s, hoof and mouth disease in western Canada in the 1950s, or Bovine Spongiform Encephalitis ("mad cow disease") in Britain in the 1990s, have in common an extremely rapid diffusion. What is most puzzling about the Black Death of the fourteenth century is its very rapid dissemination, a quality more characteristic of a cattle disease than a rodent-disseminated one.

That cattle were ravaged by these epidemics is certain. The question remains whether a natural anthrax mutant could be communicated to humans. The answer appears to be in the affirmative. Eating tainted meat from sick herds of cattle was a form of transmission to humans just as eating chimpanzees in what is today the Republic of Congo is believed by scientists to have started the AIDS disease in East Africa in the 1930s. The "mad cow" disease that killed about seventy in Britain in the 1990s was transmitted to humans by eating tainted meat.

But in 1995 David Herlihy rejected Twiggs's thesis on the grounds there were no known outbreaks of anthrax among British cattle in the mid–fourteenth century.

The response to Herlihy's dismissal came in 1998 from Edward I. Thompson of the University of Toronto. He cited a report in 1989 of an archeological excavation done at Soutra, seventeen miles southeast of Edinburgh, where a mass grave for Black Death victims was located outside a medieval hospital. The excavation yielded three anthrax spores from a cesspool into which human waste was discharged.

Thompson also cited ten medieval abbeys or priories whose cattle herds were known to be diseased. To that he added evidence, drawn from a contemporary document—

the smoking gun—from the decade or so before the Black Death of meat from cattle "dead of murrain" (meaning cattle disease) being sold in local markets.

Anthrax spores buried in the ground remain active for half a century or more as extremely toxic for humans. During World War II both German and Allied biomedical scientists developed anthrax to use in germ warfare. It was employed by neither side in the end, but the Allies tested their variety on an island off the Scottish coast. Fifty years after the war live spores buried in the ground there were discovered and the inhabitants of the island had to be evacuated.

To these known facts and Thompson's excellent work may be added a paper by Gunnar Karlsson of the University of Iceland. The island was hit by plague in the fourteenth century, but there appear to have been no rats in Iceland before the seventeenth century. The argument against Karlsson's no rats in Iceland thesis would run like this: Rats carrying plague-ridden fleas got off the boats from Norway or England but immediately died of the cold weather. The fleas then migrated to the nearest warm bodies, namely humans. Possible, but farfetched.

Thompson's conclusion that "bubonic plague and anthrax probably coexisted during the fourteenth century" is the best that science can currently provide.

If medieval physicians did fail to differentiate two separate kinds of plague during the Black Death, that should not surprise us: The scientific method had not yet been invented. When faced with a problem, people in the Middle Ages found the solution through diachronic (as opposed to synchronic) analysis. The diachronic is the historical narrative, horizontally developing through time: "Tell me a story." With their fervent historical imagination, medieval people were very good at giving diachronic explanations for the outbreak of bubonic plague in Europe and the Mediterranean region in the 1340s. One account was that the pandemic began with climatic disasters and earthquake in China, causing floods, from which came disease that moved westward.

Most medieval diachrony held that seaport towns in the Crimea sometime in the late thirteenth or early fourteenth century were so stricken by plague that when a town was besieged by an unfriendly army, the townsmen heaved corpses infected with plague over the walls at the enemy encampment—bubonic germ warfare missiles. There is a question whether plague-ridden corpses can actually communicate the disease. Most scientists say no, as a matter of fact, but that judgment is not conclusive.

Today, however, we have scientific—or synchronic—means of analysis. Science has indeed told us much more about this plague of six hundred years ago than the people living then knew themselves. It has not answered every question, but it has yielded surprising specifics.

For example, a team of research doctors at the Division of Infectious Diseases, St. Joseph's Hospital and Medical Center, Paterson, New Jersey, reported in 1997:

"Plague is a zoonotic infection caused by Yersina pestis. . . . Animal reservoirs [carriers] include rodents, rabbits, and occasionally larger animals. Cats become ill and have spread [the] disease to man. . . . Flea bites commonly spread disease to man. Person to person spread has not been a recent feature until the purported outbreak of plague and plague pneumonia in India in 1994. Other factors that increase risk of infection in endemic areas are occupation—veterinarians and assistants, pet ownership, direct animal-reservoir contact especially during the hunting season, living in households with [a disease] case, and, mild winters, cool, moist springs, and early summers."

In 1996 a team at the Laboratory of Microbial Structure and Function, working with funding from the National Institute of Allergy and Infectious Diseases, National Health Institutes, were actually able to explain what went on within a plague-infected flea: "Yersinia pestis, the cause of bubonic plague, is transmitted by the bites of infected fleas. Biological transmission of plague depends on blockage of the foregut of the flea by a mass of plague bacilli. Blockage was found to be dependent on the hermin storage (hms) locus. Yersinia pestis hms mutants established long-term infection of the flea's midgut but failed to colonize the proventriculus, the site in the foregut where blockage normally develops. Thus the hms locus markedly alters the course of Y. pestis infection in its insect vector, leading to a change in blood-feeding behavior and to efficient transmission of plague."

This is something medieval medicine did not know: the inner life of a sick flea.

More important, perhaps, is that if it is identified early enough, the plague can be cured by science today. For many years I told my students at New York University that if they are taking a shower in the college gymnasium and the person in the next stall emerges with black welts under the armpits and in the groin (the infamous plague buboes) they should dress and leave immediately. And if a rat runs by as well as the buboe-marred student, don't even bother to dress. Wrap a towel around your body and head for the nearest exit. Like most things I said in class, this got a big laugh; they didn't believe me. But I was serious.

What is truly frightening, as reported in the *Royal Geographical Magazine* in 1998, is that around the world strains of infectious disease, especially tuberculosis and meningitis, not excluding bubonic plague, that are newly resistant to antibiotics are turning up. Recently a superstrength level of antibiotics has been announced to deal with the problem. It is man versus microbe in a continuing, escalating battle.

But the chance of dying of bubonic plague in the U.S.A. today is much less than the chance of being killed in an airplane crash. Don't worry—yet. And there is more good news: the Black Death may also have protected you against the current AIDS scourge.

As part of the intense study of human genetic development related to the genome project—the mapping of the total genetic structure of human beings—a team of six scientists at the National Cancer Institute's Laboratory of Genetic Diversity in 1997 made an exciting announcement. They discovered that a genetic mutant that "occurred in the order of 4,000 years ago" gave today's human carrier of this mutant (called CCR5) immunity against HIV and therefore AIDS.

One of six signatories to the accompanying article in the *American Journal of Human Genetics*, Stephen J. O'Brien, in the following year in the same journal revealed an even more startling follow-up discovery. The mutant CCR5 could in fact, said O'Brien, be traced back to only seven hundred years ago. At that time a "historic strong selective event involving a pathogen that like HIV-1 utilizes CCR5" established an immunity "in ancestral Caucasian populations." Eighteen scientists from around the world affirmed O'Brien's hypothesis.

The event he described, of course, could only be the Black Death. There is thus—if O'Brien is correct—a genetic relationship between the Black Death and AIDS. If you are descended from a Caucasian who contracted the plague of the mid–fourteenth century and that ancestor survived, you may have complete immunity to HIV/AIDS. And it is believed that up to 15 percent of the Caucasian population could fall into this lucky category.

୧ଆଙ୨

Whatever the plague was—or continues to be—it still remains with us in its original form. Since the advent of antibiotics in the 1940s bubonic plague is very rare in the U.S.A., although there are still substantial outbreaks in eastern Asia, especially India. But in the 1980s there were three documented cases in the hill country of eastern California. One woman came down with the bubonic plague after she ran over a squirrel with a power mower. It is likely that the disease entered California at some port on a rodent traveling on a ship from eastern Asia.

Medieval doctors did not know about the bacillus parasite carried on the backs of rodents. It was assumed that it was spread through the air—as a miasma—from person to person. This induced healthy people to flee cities to isolated country retreats, as was done by most of the English royal family and as described for Florence in Boccaccio's *Decameron*. Modern medicine believes that plague can be spread by saliva from an ill person to a healthy one at the pneumonic stage, so medieval physicians were not entirely wrong. Rodents, however, are clearly central to the problem.

Since rodents were common in the Middle Ages, even in the residence of the affluent, escape was not easy. However, the impact on the rich (from English statistics) seems to have been not more than 25 percent mortality, while for peasants—including parish priests—mortality averaged in the low 40s and in some places as high as 50 percent. Given the difficulty that medieval towns had getting rid of sewage on which rodents feasted, the urban mortality was also around 40 percent, at least for commoners.

Rich urban people fled to country retreats and may have done better. But archbishops, great lords, and wealthy merchants also fell to the plague: it was a democratic disease.

Since medieval physicians were convinced the plague was airborne—a miasma—they induced a change of lifestyle. Windows must remain closed and covered—for the affluent, with thick tapestries. The Black Death did wonders for augmenting the market for the tapestry makers in Belgium and northern France. The magnificent late medieval tapestries in the Cloisters in New York City and the Cluny Museum in Paris were therefore functional as well as decorative.

Like so much of medieval art, tapestry making originated in the Middle East and spread to Europe through Byzantium. In the West it was taken up by the monks who were already the primary providers of medieval artwork. But the very large tapestries now called for to block the pestilence from entering the homes of affluent people required workshops of imposing size and droves of highly organized and well-paid workers. Mere heavy window coverings were not enough for the rich, who wanted elaborate embroidered narratives of their favorite scenes from popular romances.

The weaver guilds in Flanders and northern France responded quickly to the demand.

Frequent bathing was proscribed as dangerous by the medical profession: You opened your pores to the airborne disease. Europe entered the pungent no-bath era, which lasted until the disappearance of the plague in the mid–eighteenth century. Even Napoleon Bonaparte rarely bathed; instead he had a massage with French cologne each morning, a lifestyle common to the European nobility by 1400 and a legacy of the Black Death and medieval medicine.

Inevitably medieval physicians attributed the onset of the disease to God's punishment for sin and to bad astrological conjunctions involving the feared planet Saturn. The king of France appointed a commission of University of Paris professors to account for the Black Death. The professors soberly blamed the medieval catastrophe on the astrological place of Saturn in the house of Jupiter.

The most immediate problem caused by the mortality of the Black Death, whatever its clinical components may have been, was how to give the dead decent Christian burial. Since parish clergy were hit as hard as any group in society by the pestilence, there was a shortage of priests to administer last rites and preside over funeral services. Nor could gravediggers keep up with mortality. Inevitably the solution was to engage in mass burials.

All over Western Europe commoners were buried in mass graves with bodies stacked horizontally five layers deep.

Archeologists have discovered such layered mass graves in many places, including central London. Since the earth covering the mass graves was thin, the stench rising from the cemeteries was initially unbearable.

In England the catastrophic rate of mortality did not immediately produce a severe labor shortage in countryside and town. England in the early 1340s was still so heavily overpopulated that peasants were available to take over vacant farms and empty slots for estate workers. The diminished population of cities and towns was replaced from the long waiting list of populations seeking entry to alleged privileged urban life.

But by the next generation, in the 1370s, the Black Death had caused a critical labor shortage, especially in rural areas. Peasants took advantage of the labor market operating in their favor to demand steep increases in wages from landlords. The aristocracy and gentry responded by using Parliament to force through laws holding down workers' wages against the inflationary labor market.

This governmental intervention was a prime cause of the outbreak of the Peasants' Revolt of 1381 in eastern England, the greatest proletarian rising before the eighteenth century. Urged on by radical clerics, the rebellious peasants came close to bringing down the government and establishing a Christian socialist regime.

After the Black Death had raged for more than a year some cities came up with a preventive measure that did some good—

or so it was claimed—through the strict quarantining of areas of the city where the incidence of pestilence was heavy. What they really needed to do was quarantine rodents, but human quarantine measures apparently were effective in some towns— or so the town officials, having already disturbed social life, asserted, to make their quarantine policy appear effective.

They should not be blamed for doing so.

The level of mortality in the Black Death was so high and so sudden that—until germ warfare on a large scale occurs—to find a modern parallel we must look more toward a nuclear war than a pandemic. The plague shook the wealthy, relatively well-populated, confident, even arrogant society of mid-fourteenth-century Western Europe to its foundations.

The survivors of the biomedical holocaust were at first too stunned and confused to do more than augment religious exercises. But slowly it was realized that institutions and the populace would be deeply affected by the great biomedical devastation and sudden severe shrinkage of the population. At various levels of society there were challenges to the old order and there were adjustments to be made to a drastically affected world. The pestilence deeply affected individual and family behavior and consciousness. It put severe strains on the social, political, and economic systems. It threatened the stability and viability of civilization. It was as if a neutron bomb had been detonated. Nothing like this has happened before or since in the recorded history of mankind, and the men and women of the fourteenth century would never be the same.

PART II

People

〜〰〜

CHAPTER THREE

Bordeaux Is Burning

BORDEAUX IS SITUATED ON THE broad Gironde estuary on the west coast of France. It was one of the great port cities of the Middle Ages. It was part of a territory in western France called Gascony that was owned by the English royal family.

In August of 1348 Bordeaux was visited with much triumphal pomp by the daughter of the English king, who was on her way to Spain to marry the heir to the throne of Castile, the largest of the Iberian kingdoms.

The baggage that fifteen-year-old Princess Joan carried with her as her ships sailed up the estuary to the harbor at Bordeaux was not only material. It was also historical. Centuries, indeed millennia, of strife and tradition of enterprise and culture had made the little princess what she was, and how she was regarded by awestruck merchants who greeted her at quayside in Bordeaux.

She was a top-drawer white girl, a European princess, with all that this status and image meant in fourteenth-century world society. She was a product of Caucasian mastery that in the late fifteenth century would explode overseas, first to Africa then to eastern Asia and the Americas where it would bring European civilization and all that it meant in terms of power, exploitation, learning, wealth, and misery to other branches of mankind.

Western Europe of A.D. 1340 was a politically pluralistic society—that is, government, law, and taxation, were actually administered by a patchwork quilt of country and city holders of power. But everyone held in great honor and esteem the kings and their royal families.

Kings were regarded as anointed by the Lord, and holders of divine power. They were the echoes and images of the authority wielded by the antique Roman emperors. They were the perpetuators and heirs of the old German chieftains who were much appreciated for their personal strength and valor and for their distribution of gold rings and other booty.

Most kings filled these awesome roles weakly and uneasily, like third-rate actors playing Hamlet on road circuit in the boondocks. But occasionally there were ambitious and energetic kings thrown up unexpectedly by lottery of birth, and with the assistance of good training and education they would push hard to mobilize the Christian, Roman, and Germanic traditions of kingship.

Invariably this led to fighting and misery for some people, but often the kings made an impact and they gained renewed admiration, respect, and fear for the royal family among the common people and even a certain caution and good manners from the great and very rich landholders.

Steeped in art, poetry, and music, and decorated with expensive plate armor and the latest silk clothes in the Milanese or Parisian mode, this aristocratic, very rich people's respect for the king and his family became the behavioral pattern and sensibility called chivalry. The word originally meant horsemanship. Owning good riding horses was an expensive proposition, normally confined to the nobility or their high bourgeois imitators. Even today on the Main Line outside Philadelphia or in Fairfield County, Connecticut, owning a show horse is a sign of wealth and good manners.

Some kings gained much popular prestige and chivalric admiration and used it to their advantage by hiring clerks and lawyers to construct a formidable centralized bureaucracy for the purpose mainly of improved taxation mechanisms. Usually the effectiveness of such centralized government sharply attenuated if the great king's successor was a weakling, but shards of the original administrative structure could run on for centuries with little rationale and control.

Such was the history of medieval monarchy and royal dynasties. Yet whatever the king was actually doing in relation to the people (all the way from nothing to strong leadership to tyranny), members of his immediate family, his wife and children, were held in greatest awe and honor by sentient strata of society.

This meant not only the nobility and gentry but the

merchants, capitalists, bankers, and prominent tradesmen in town. In the depth of their crowded houses, surrounded by furniture, art, and servants, the high bourgeoisie might think or even express controversial democratic thoughts. Occasionally someone in the merchant entourage would write down these ideas, nearly always in guarded fabular or metaphorical form. And once in a long while, if a monarchy had suffered terrible battlefield losses that greatly diminished prestige, a group of capitalists might assert political urban autonomy.

But normally the rich townsmen kept their radical thoughts to themselves and bent the knee as they greeted a scion of the royal family with trumpets and lavish gifts.

From the long quays of the port of Bordeaux, ships customarily departed for England carrying barrels of red wine from vineyards that had in the twelfth century belonged to the famous duchess Eleanor of Aquitaine. By her second marriage (after divorce, technically annulment of marriage, from the king of France) Eleanor had in 1152 married the nineteen-year-old stud Henry Plantagenet, count of Anjou, ten years Eleanor's junior. Eleanor's wealth and political influence helped Henry Plantagenet to assert his hereditary claim to the English throne, and he ruled as the great monarch Henry II until 1189.

Queen Eleanor lived on into the thirteenth century, long

enough to see two of her four sons by Henry Plantagenet (she had two daughters by the king of France before her divorce) sit on the English throne as Richard I the Lion-Hearted and the manic-depressive John. The latter came to be called John Lackland because he lost the ancestral territories of the royal family in Normandy and Anjou to the king of France and some of Eleanor of Aquitaine's vast domain in western France as well.

During the long reign of Henry III, John's pious, feckless and cowardly son, more of Eleanor's patrimony was lost to the ever-expanding French monarchy, including her capital city of Poitiers.

But in 1348, in the midst of a war that Henry III's great-grandson Edward III had begun in 1342 (it is known to us as the Hundred Years War) to recover the ancient French Plantagenet lands, Bordeaux was still a great port for shipment of wine to England. It was the principal city of the English-ruled territory called Gascony, which stretched narrowly (never more than a hundred miles wide) along the French coast from Brittany in the north to the foothills of the Pyrenees in the south.

The English nobility and gentry, which drank many thousands of gallons of Queen Eleanor's red Bordeaux wine each year, came to call it claret, which means clear, that is fresh and cool wine. Labels from Gascon vineyards within fifty miles of Bordeaux still stand out in the French wine section of our liquor stores today—names like Graves and St.

Emilion. But the prestige label is Château Lafite Rothschild, from the centerpoint of Eleanor of Aquitaine's vineyard. The Jewish banking family bought these vineyards in the second half of the nineteenth century. Today Lafite is so prized that it is sold normally only at auctions, at a minimum of three hundred dollars a bottle.

So does the heritage of medieval history and the grandeur of the Plantagenets flow down the throats of the rich today, just as it was relished by the French-speaking English nobility in the fourteenth century.

It was a glorious scene in early August 1348 as four English ships with sails set and banners flying sailed down the Gironde estuary and docked at the port of Bordeaux. The lead ship carried Princess Joan, the daughter of King Edward III, on her way through Gascony southward to Spain, where she was engaged to marry Prince Pedro, heir to the kingdom of Castile.

Castile, centrally located in the Iberian peninsula, was renowned for its wool and grain and its fierce nobility, who had sharpened their military skills in two centuries of fighting and pushing back the Muslim Arabs, driving them into Grenada, a small redoubt in the southwest corner of the peninsula. Here the Arabs were finally overcome in 1492 after the union through dynastic marriage of the two richest Iberian kingdoms, Castile and Aragon.

At the battle of Crecy in 1346, two years before Princess Joan landed at Bordeaux and retired to the royal family's

Château de l'Ombriere overlooking the estuary of the Gironde and the main port areas of Bordeaux, Edward III had won a devastating victory over the French king and nobility. Edward had surprised aristocratic Europe by showing that the days of aristocratic prowess through cavalry charges by heavily armored knights were numbered.

The English army had fought mainly as infantry, protected by leather and thin armored pieces. They were well-trained and generously paid peasants who fought as pikemen, wielding big globs of iron to break the charge of the French cavalry, and as bowmen. From fighting the recalcitrant Welsh in the late thirteenth century the English had also learned that peasants using longbows could create havoc on a cavalry charge by firing showers of metal-tipped wooden arrows in rapid waves, maiming, killing, and terrifying the enemy's horses and sometimes even penetrating some knights in plated armor.

The French had mercenaries wielding crossbows, which fired a devastatingly heavy metal bolt often demonstrated today in Hollywood horror movies. But the crossbow range was short, only thirty yards, and once a French crossbowman had shot his bolt (the origin of the phrase) he would need an assistant and a half-hour to reload. The English longbows were rapid-fire weapons dispatching death and confusion two hundred yards upfield upon the advancing French knights and their vulnerable horses.

England in 1346 had one-third the population of France

and at most half the gross domestic product of the continental kingdom. Edward was only able to mount powerful armies against the king in Paris because of a much superior English tax system. This generated resources enough to fight and win great battles against the French but not enough to achieve Edward's ultimate goal of making himself king of France as well as England.

Yet for twelve decades the English held on to large stretches of the western third of France. Their mercenary bands of "free companies," consisting of gentry, peasants, occasional lords, and not a few professional criminals, ranged the French countryside, burning, looting, kidnapping, and raping. Finally in the 1440s the English Parliament got tired of paying for this Holocaust and the French—according to doubtful national legend rallied by a visionary peasant girl, Joan of Arc—defeated the marauding English and drove most of them out. The English kept only one French city, the port of Calais, until the mid–sixteenth century, and they never again made aggressive war on the European continent, preferring to create instead an empire overseas.

Even then, not all of the English army went home. Some made their peace with the king in Paris and settled down in Gascony, which had been their homeland for a century. Among these was Captain Hennessy, an Irish mercenary fighting in the English army. He settled down near Bordeaux and learned from Benedictine and Carthusian monks how to make brandy. The Hennessy label on cognac is almost as prestigious today as the Rothschild label on Bordeaux wine.

Princess Joan thus landed in Bordeaux in early August 1348 at the high point of fourteenth-century Plantagenet fortunes, when there still seemed a good possibility that Edward III and his successors would sit on the Parisian throne (Edward's great-grandson Henry V, after his victory at Agincourt in 1415, again came close to achieving this goal but died young before he could work out the political and diplomatic angles).

Now that France seemed to be on the verge of being swallowed up into the domains of the awesome and restive Plantagenets, Edward looked further afield to the rich cities and fecund plains of Andalusia. He would marry his fifteen-year-old daughter Joan to the heir of Castile and eventually the Plantagenet line would prevail in Spain as in England, Wales, and France.

Edward III always couched his imperial ambitions in the language of hoary dynastic claims and refined aristocratic honor. He was the founder of the super-elite aristocratic Order of the Garter. His propagandists in letters and art presented him as a King Arthur incarnate, as the embodiment of European chivalry, as the exemplar of virtuous noble temperament, as the purest refinement of Christian militarism, a gentleman's gentleman.

Edward III in fact was an avaricious and sadistic thug who aimed to conquer much of Western Europe, from Flanders in the north (what is now called Belgium) to the Strait of Gibraltar at the tip of Spain in the south.

From the heated loins of Henry Plantagenet and Eleanor of Aquitaine had sprung a genetic order of fighting royal monsters. Edward III was the epitome of this devilish breed. He was personally brave, a skillful general, a good organizer. Edward III's eldest son, Edward the Black Prince (so-called for his arms and his heart), was the exact copy of his ruthless, devious, and greedy father.

As Edward III aged (he didn't die—from gonorrhea—until 1377), the Black Prince took over leadership of the English continental armies, laying waste to huge parts of France and Spain. The Black Prince, overcome by malaria he contracted fighting in Castile, died a few years before the horrible old man, who was clutching his venereal mistress to the end.

Did contemporaries think of Edward III as an evil scourge? Plenty of French peasants did, but among the articulate and literate classes, aside from a handful of radical friars, he was not even considered a tyrant. That term, derived from Roman writers, was reserved for an absolute monarch who ruled without consent of the people. By that definition, Edward III was no tyrant.

Almost yearly he met in Parliament with the lay and ecclesiastical magnates, and with representatives of the gentry and the merchant class. The king could not impose his more lucrative taxes without their consent, and usually he got what he wanted. Some important legislation was also presented to him by Parliament, and he usually approved it.

Probably Edward III would have preferred to rule by fiat and dispense with Parliament, but constitutional development under his grandfather and father in the previous ninety years prevented him from doing so.

That today we may look back on the English king of the fourteenth century as a kind of destructive and merciless force, while to nearly all articulate and literate contemporaries he was a constitutional king and very model of chivalry and aristocratic honor, illuminates a gap between our world and fourteenth-century Europe.

Fourteenth-century people lacked the moral categories that could transcend traditional political and social roles. They lacked a critical value system that judged rulers by consequences and not the formal categories in which their behavior was structured.

There were plenty of passages in the biblical prophets and Gospels to condemn the Plantagenets' savage behavior in the Hundred Years War, and there were a few churchmen at the margins painfully aware of this. But the magnates of the Church—the pope, cardinals, bishops, and abbots— were too enmeshed in the prevailing political and social nexus to assess critically the behavior of the English crown. Themselves from the aristocracy or gentry, they accepted current society and its values.

Edward III and the Black Prince were regarded as divinely ordained forces of nature like the sun and the wind. In person they were admirable gentlemen, superior human beings. As officeholders they were capable of distributing generous patronage. They were to be obeyed and eulogized, not criticized or condemned. It was natural for Edward III to press his marginal claims to the French throne, inevitable that hundreds of thousands of commoners would suffer in the consequent wars.

Again and again the pope sent emissaries to make peace between the English and French kings; the popes and legates were rebuffed, especially by the English government. The pope was living in exile from Rome in Avignon on the Rhone river and was regarded as a French puppet. This futile peace-making gesture sufficed to calm ecclesiastical guilt. The moral and intellectual conscience of the church was suffocated and stilled within the structure of wealth and power.

But the coming marriage of Princess Joan and Prince Pedro loomed as a great event in every respect, political, religious, and diplomatic. Joan rested at the royal castle overlooking the port of Bordeaux before proceeding southward overland through Gascony to Castile. The marriage had the pope's eager blessing, although the dynastic union in the long run would lead to yet more savage conflict in both France and the Iberian peninsula.

Religious authorities, whether priests or rabbis, are always in the front rank of celebrants of the marriage of the scions of rich families. It is and was an appearance they relish making, and not just because of the succulent gifts that they will receive from the families involved. They are happy to perform ceremonies in festive and lavishly decorated surroundings that the rich and powerful own.

Fifteen was not unusually early for a royal princess like Joan—or the daughter of any rich man—to marry in the fourteenth century. Whenever a girl of the more affluent classes, from Jane Austen–like middle-class gentry or merchant all the way up the social ladder to the royal family, reached the age of menstruation, she had only two life prospects before her: marriage or the nunnery. And a royal princess was too valuable a commodity in political machination and diplomacy to be wasted as a decaying virgin in a convent.

Women in the Middle Ages had an even shorter life expectancy than men as long as they continued to produce children. Their frequent pregnancies and childbirths commonly led to death by thirty from some obstetrical or gynecological complication. Medieval princes, noblemen, and gentlemen tended to have serial marriages because of the Russian roulette of pregnancy and childbirth imposed by crude medical science upon their wives.

The male rich and powerful were often on their third or fourth marriage by the time they died before their forty-fifth birthday—of natural causes, infectious disease, or heart attacks from a very high-cholesterol diet, if they were not struck down earlier in battle or brawl. If a queen or other rich woman did not get pregnant and give birth she was shunted off to a nunnery and a new, younger, and perhaps more fruitful wife was chosen. It was only the wife who was considered to be infertile, never the husband.

Since menopause then came around age thirty, the wife's main job as breeder was fulfilled when childbirth crisis struck her down around that age. When a Sicilian princess around 1200 produced a son at age forty it was regarded as a miracle, with parallels to the virgin birth of Christ—so said the court propagandists of the emperor Frederick II, the celebrants of this unheard-of event.

A modern actuary would have given the menstruating fifteen-year-old princess Joan just another ten years to live. This is why the premature death of young women of royal and noble families generated modest grief.

This was not the first time that princess Joan had been abroad. As a five-year-old she had been taken on the king's trip to meet the German emperor at Coblenz. No effort was spared now in the preparation for Joan's wedding venture, although as was usual with royal weddings Edward III shifted as much of the cost as possible to his long-suffering tax-paying subjects. From December 1347 several ships had been commandeered along the south coast of England for the voyage to Bordeaux.

Three flunkies of the royal household were dispatched to purvey (that is, extort) food from the south coastal county of Devon (later the scene of most of Thomas Hardy's novels). Baron Robert Bourchier, a man of substance and high visibility and former royal chancellor, was to head the diplomatic delegation accompanying the princess. The king was too busy organizing and fighting his French war to accompany his daughter. Her mother, Queen Philippa of Hainault, never traveled abroad except to visit her homeland in the Low Countries. Bourchier had served as the head of the royal administration in 1340–41. He was an accomplished diplomat and soldier and had fought with distinction at Crecy.

Another member of the princess's entourage was Andrew Ullford, an Oxford doctor of civil (Roman) law, who held a high position in the cathedral of York. Ullford was also an experienced diplomat. The king sent these important diplomats to assure that a treaty would be drawn up before the marriage of Joan and Pedro assuring that any son born to this union would succeed to the throne of Castile regardless of any subsequent marriage by Pedro.

The princess's spiritual needs were to be addressed by a prominent priest of Bordeaux Cathedral, Gerald de Podio. Then there was the minstrel. Prince Pedro had dispatched to England his favorite court minstrel, Gracias de Gyvill, to entertain Pedro's betrothed with the songs of the land of which she was to be queen, a charming idea.

Princess Joan was also to be accompanied by one hundred formidable English bowmen, some veterans of the Battle of Crecy. They were not just a ceremonial bodyguard. Traveling through long stretches of thinly populated southern Gascony had its risks and dangers from criminals and freebooting mercenaries, perhaps even from an avenging agent of the king of France, who could only look with fear on the prospects of a Plantagenet-Castilian alliance against him—as if the Parisian king didn't have enough troubles already, not only from English armies and brigands but from the politically restless bourgeoisie of Paris, scorning the king and his nobility for their military defeats.

One English ship was needed to carry to Bordeaux the lavish clothing and other belongings of the little princess. In dressing and equipping Joan, Edward III had characteristically spared no expense, partly out of love for his daughter and partly as a display of kingly prowess and wealth toward his Spanish allies and prospective in-laws.

Over 150 meters of rakematiz, a thick imported silk woven with gold, were used to make Joan's wedding dress. She also had a suit made from ten pieces of red velvet. Two of her five corsets were made from cigaston, the heaviest silk, newly in fashion at Edward's court, and woven with gold patterns of stars, crescents, and diamonds. She had two sets of twenty-four buttons, each one made of silver gilt and enamel.

There were two elaborate dresses, called ghitas, with an inbuilt corset. Both of these ghitas were also made of rakematiz, one in green, one in dark brown. The green was embroidered all over in gold with images of rose arbors, wild animals, and wild men. The brown had a base of powdered gold onto which was set a pattern of repeating circles, each enclosing a recumbent lion, symbolic of monarchy, each embroidered in bright colored silk and metal threads. Good taste was not a quality of the English monarchy then or now.

Princess Joan traveled with a portable chapel so she could enjoy Catholic services and sacraments without having to use local churches and encounter commoners in her travels southward. The lavish private chapel featured a couch decorated with fighting dragons and a border of vines, powdered throughout with gold Byzantine coins. The vestment cloth meant to cover an altar table was decorated with serpents and dragons.

Among the silver vessels was an incense burner valued at half a million dollars in today's money and a silver chalice also of similar value. It was a long way from the Sermon on the Mount.

The mayor of Bordeaux in August 1348, Raymond de Bisquale, greeted the princess and her large entourage and escorted them to the Château de l'Ombriere, the old Plantagenet castle on the estuary overlooking the port. Mayor Raymond commented to Joan and her three leading officials, the former royal chancellor Robert Bouchier; the diplomatic lawyer Andrew Ullford; and the local cathedral priest, Gerald de Podio, that plague was causing trouble in Bordeaux.

Hundreds of cadavers with the dreadful buboes, black welts under the armpits and around the groin, were piling up in the streets and on the docks. The stench was almost unbearable. But medieval lords were used to bad smells. They took care of them by holding silk handkerchiefs drenched in perfume to their noses.

The royal entourage ignored the mayor's warning about the plague. The princess and her companions proceeded to settle comfortably into the royal château even though it was located in a dangerous place near docks swarming with plague-carrying rats.

The royal castle was situated close to the main port area by the estuary of the Gironde on which Bordeaux is located. Like very rich people in all times and places, the Plantagenets built their establishments on prime real estate located on the water if it was at all feasible. But this old and expensive part of town was particularly affected by the plague, which tended to travel along the river trade routes, often in bales of cloth, which could house rats carrying infected fleas ready to migrate onto warm human bodies. At Narbonne, at the southern end of Gascony near the Spanish border, and a town which the princess and her entourage would certainly have stayed before they crossed into Spain, the plague similarly began among the dyers working in factories along the river.

In Bordeaux, among the bales of wool and cloth that stood on the quays next to the huge barrels of red wine destined for England, rats scurried around carrying the infectious bacilli.

Very soon the fifteen-year-old princess watched as her companions fell sick and died of the plague. A year later her father the king and her brother Edward the Black Prince fled with their entourages to distant estates and castles in England. Whether by this flight into remote areas or by sheer luck, they saved themselves. It did not occur to the little princess and her upscale advisors to get out of town. They stayed at the very center of the pestilence raging in Bordeaux, with devastating consequences.

The royal château overlooking the water became a charnel house of horror. Robert Bourchier, the chancellor who had survived the Battle of Crecy, was one of the first to be struck down by the plague. He died on August 20. On September 2 Princess Joan died, a deep personal and major diplomatic setback for Edward III.

Andrew Ullford the lawyer and diplomat was not affected by the plague (he in fact resumed a busy diplomatic career and died ten years later on a mission to the Roman curia in Avignon). Now Andrew took off for England and on October 1 reported to the king what had occurred. The princess was dead of the plague.

King Edward immediately sent a letter to King Alfonso of Castile terminating the marriage arrangements and concluded with traditional and formal piety: "We have placed our trust in God and our life between His hands where He held it closely through many dangers." Edward had already married off his other daughter to an English earl. He had no more female progeny with whom to further pursue the marriage and strategic alliance with Castile.

Yet Castile remained a ghostly, beckoning specter in the Plantagenet consciousness. Princess Joan's brothers the Black Prince and John of Gaunt were each to make noisy and bloody expeditions into Spain in the following decades to try to carve out spheres of English dominance there by direct and forceful means. Neither of these expeditions led to any permanent consequences. The Spaniards were too good as fighters and politically more savvy (or treacherous) than even the Plantagenet dynasty.

The fate of Pedro's minstrel sent to England to saturate Joan with Spanish songs is unknown. Probably he died in Bordeaux of the plague. Robert Bourchier's body was brought back to England, and he was buried at a small monastery he had endowed in the county of Essex.

There is no record of Princess Joan's body being returned to London, nor any account of her funeral, which would have been a well-appreciated event. This strange lacuna was not due to Edward's negligence. No medieval king allowed his daughter's body to disappear abroad unburied.

On October 25, 1348, the king commissioned a northern ecclesiastical lord, the bishop of Carlisle, to go to dangerous plague-ridden Bordeaux and bring back the little princess's body for burial in London. The bishop was generously allocated five marks (about two thousand dollars in our money) a day in expenses for this task. Undoubtedly Edward III overpaid the good bishop because of the health risk involved.

It is just possible that the bishop chickened out and never dared go to plague-infested Bordeaux, which would have been a very rash default on his part. More likely Carlisle went to Bordeaux but could not retrieve Joan's body because it had been turned to anonymous ashes in the general conflagration that had engulfed the port area.

In Bordeaux the port area was so badly affected by plague (everyone died there of this great mortality, remarked a contemporary French chronicler) that Mayor Raymond de Bisquale decided to set fire to the port area in the hope of stopping the spread of the epidemic. The flames and smoke of the burning port could be seen for miles.

The physicians the mayor consulted, if they did not blame the whole biomedical disaster on insalubrious astrological signs, would have told the mayor that the plague passed through the polluted air from the mouth of a sick to a healthy person.

But the fire in the port got out of control and destroyed some valuable housing quite a distance from the port area, as we know from a later lawsuit. The flames engulfed the royal château on the estuary and, in all likelihood, took Joan's remains with it.

This sad end for the royal princess heightened the king's grief. In his letter to King Alfonso of Castile on September 15, 1348, Edward referred to "the intense bitterness of heart" at what had happened. The princess was a martyred angel looking down from Heaven to protect the king and the royal family: "We give thanks to God that one of our own family free from all stain [she was after all a virgin] whom we have loved with our life has been sent ahead to Heaven to reign among the choir of virgins, where she can

gladly intercede for our offenses before God Himself"
(translated by R. Horrox).

Considering that Edward III with his son the Black
Prince had laid waste to about 25 percent of the present-day
western third of France and caused the deaths through violence and famine of hundreds of thousands of noncombat-
ant peasants and urban workers, Joan the angelic virgin had
her work cut out for her.

Mayor Raymond de Bisquale of Bordeaux, though he
did not share the Plantagenet success in politics, was luckier
in health and survived the plague. But its effects on this rich
port city and the surrounding lush agricultural areas, including the vineyards, was severe.

The banlieue (surrounding territory) of Bordeaux was
devastated by the plague, compounding serious damage already inflicted by ten years of fighting in the Hundred Years
War. In February 1365, the chapter of the cathedral of St.
Andrew asked to recover its land tenements in the village of
St. Julien, just south of the town walls, because they had
been vacant for twenty years or more. The tenants were dead
and no one had come forward to lay claim to the holdings or
pay rent. The houses had fallen into disrepair and were deserted and abandoned.

The cathedral dean and chapter did not know whether
any heirs to the properties survived. As a result proclama-
tions were read four times in the largest church and the fifteen parochial chapels of Bordeaux calling on the heirs to

come forward. On February 29, no claim having been presented, the tenements were held to have reverted to the hands of the dean and chapter.

The accounts of the Archbishopric of Bordeaux revealed other cases of vacancy and decayed rents. In 1356 at la Souys, near Floirac, just across the Gironde from St. Julien, the vineyards lay abandoned. A mile to the north at Lormont thirteen holdings lay untenanted in 1361. In the same year there were seven deserted holdings at Pessac, which lay just beyond St. Julien and its superb vineyards.

Initially precipitated by the war and exacerbated by the plague, the crisis in the wine trade, the major economic staple of Bordeaux, ran alongside this devastation of the countryside. In the first half of the fourteenth century between 725 and 1,360 ships a year set out for England carrying Gascon wine. Between October 8, 1349, and August 27, 1350, only 141 ships sailed.

Although wine exports did not regain their preplague levels, demand in the English market did not fall significantly, with the result that Gascony enjoyed the partial compensation of a significant rise in price. Although it fell back later, it generally stood at one-third above its preplague level.

The revival of the wine trade was accompanied by a process of rural reconstruction near Bordeaux, which began in 1355, intensified from 1363, and reached its peak in 1368. The Abbey of La Sauve made a systematic effort to repopulate its estates with peasants. As well as drawing on local families, immigrants arrived from the neighboring counties, and as far away as Brittany and Spain.

The new tenancies incorporated conditions for the restoration of houses over a period of three or four years and for the customary maintenance of old vines and planting of new ones within four years. The resumption of its lands at St. Julien by the dean and chapter of St. Andrew in 1365 was a necessary preliminary to the process of reletting and reconstruction. The dean and chapter were making equivalent efforts in Graves. Similarly, in another prime wine-producing village, between 1366 and 1372, the chapter of St. Seurin and the seigneur of Castelnau relet abandoned holdings, imposing the condition that the new tenants replant the vines within six years.

The death of the princess had brought home to the royal family back in England the seriousness of the threat presented by the plague. Peasants dying in manorial villages, mass graves in London and other cities—these portents initially alarmed the Plantagenet family and court. The devastation in Bordeaux, especially the death of a royal princess critically important in international politics, raised a grim spectral threat.

The court escaped comparatively lightly from the plague of 1349. The king and his eldest son the Black Prince spent the summer months on the royal manors in southwest England well away from the principal centers of population and in an area of the country where the mortality was already passing its peak. But although most of the close companions of the royal family were spared, astonishingly, eventually there had to be an emotional impact that the mass mortality imposed on the royal family and the court. One of the best-recorded responses is the arrangements made for the burial of the dead. In 1349 Sir Walter Manny, one of Edward III's great war captains, purchased a mass burial ground, the Spital Croft later called New Church Haw, and now the site of the Charterhouse.

Although this may have been principally a practical and sanitary act, it had a pious aspect, reflected in the foundation of a chapel dedicated to the Holy Trinity and the Virgin Mary within the burial ground. Manny secured a license from Pope Clement VI to transform the chapel into a secular college with thirteen chaplains. However, Manny changed his mind in 1371, a year before his death, and founded a Carthusian monastery (the Charterhouse) on the site.

The Carthusians were the strictest of the religious orders, and Manny's choice may have reflected the spirit in which he made the foundation. He left the monastery two thousand pounds ($7 million today) in his will and arranged for his own burial there, further signifying the importance he attached to his plague foundation.

Edward III himself made a similar contribution to the salvation of the souls of the plague victims. In 1350 he purchased a plague cemetery close to the Tower of London, land originally acquired in the previous year by John Corey, a clerk from Holy Trinity Priory. It is probable that Corey was acting as the king's agent. Edward founded a chapel dedicated to the Virgin, to whose intercession he attributed his own escape from many perils at land and sea.

In part Edward's gratitude reflects his thoughts about the dangers of war, but it must surely also bear some relation to the fact that he had been spared by the plague. Edward's concern was principally a personal one rather than one of compassion for the mass of victims. Soon afterward, the foundation was transformed into a full-fledged Carthusian monastery of St. Mary Graces under the supervision of the old monastery at Beaulieu Regis, from which five monks were sent to man it.

Edward's original intention was to provide the new monastery with one thousand pounds sterling ($4 million) per annum. In fact Edward gave the foundation only some properties in Smithfield and a meager twenty marks a year, showing that the project had quickly slipped down his list of priorities. In 1358 he increased the annual grant to forty marks a year from the Exchequer, though he insisted on the addition of a further monk.

In 1367 he granted the monastery income from two London churches and other rents in London worth one thousand pounds a year, fulfilling his original commitment. On his deathbed Edward made further grants on the scale that he had originally promised.

The greatest lord and richest man in England outside the

immediate royal family was Henry of Grosmont, Duke of Lancaster, a much-admired veteran of the French wars. Grosmont was deeply moved by the plague. He survived the plague attack of the 1340s, only to succumb to a similar outbreak in 1361. In 1354 Grosmont set down, at the urging of his confessor, his thoughts on biomedical matters in a little *Book of the Holy Doctors*. It is a quite moving document reflective of the plague years.

Among the personal weaknesses Grosmont acknowledged and set out to rectify was a disgust at the sick, which was so extreme that they were not even spared the scraps from his table. In the duke's book, Christ the physician assisted by the Virgin Mary offers cures for the wounds of the soul infected with the seven deadly sins.

Grosmont had been alerted to, if not entirely converted to austerity. He tells us that as a young man he had been tall and fair, fond of adornments such as rings and garters and with a proclivity for dancing. When he was older he had a dancing chamber in his castle at Leicester and a personal troupe of minstrels. He liked women and preferred those of low birth because they did not criticize him for his desires.

He had the usual fondness of his class for hunting and warfare. Now that he was growing old, he still indulged in rich, spicy food with strong sauces—salmon and lamprey were his favorites and were bought for him at great cost.

He also downed large quantities of claret and openly declared that he and his companions readily lost their senses in wine. What the book shows is Grosmont's realization of the spiritual limitations of this lifestyle and the disparity between the luxury available to him and the uncomfortable lives of most of his fellow Englishmen.

The same awareness is reflected in the arrangements made for his burial, on April 14, 1361, at the monastic college that he founded at Newark. He eschewed all military display—there were no armed men or horses with trappings. Instead the hearse bore five great candles of one hundred pounds each as well as four funeral lights and was preceded by fifty poor men, half of them in blue, the rest in white, each bearing a lighted torch. The royal family attended the funeral and the Black Prince placed two cloths of gold on Grosmont's bier.

Quite apart from personal loyalty the royal family had every reason to be there. Grosmont left no male heir, and when one of his two daughters, Maude, died soon afterward, the entire vast estate passed to his other daughter, Blanche, wife of the king's younger son, John of Gaunt.

The new duke of Lancaster, John of Gaunt, was Edward III's second-eldest surviving son, after the Black Prince. When Edward died in 1377 the crown passed to young Richard II, the Black Prince's son by the love of his life, the vivacious Princess Joan of Kent.

The passing of the vast Lancastrian holdings and the ducal title to John of Gaunt destabilized the Plantagenet family, because John of Gaunt, now duke of Lancaster, could command as much property and as many soldiers as the rest of the royal family. Almost inevitably Henry of Lancaster, John of Gaunt's heir, threw out his gay cousin Richard II and seized the crown with parliamentary approval. Henry IV had Richard II taken to a bleak castle and probably starved to death.

The rest of the Plantagenet family now coalesced slowly around the duke of York, descended from Edward III's fourth son, Edmund of Langley. Thereby the stage was set for the twenty-five-year struggles between the Houses of Lancaster and York, which Victorian writers called the War of the Roses. It was the long, seemingly interminable civil war between Lancaster and York factions of the Plantagenet dynasty that made it impossible for the English to stand against the revived, Joan of Arc–inspired French monarchy in the 1440s and 1450s, which is how Bordeaux and its neighborhood and vineyards were lost.

John of Gaunt got his peculiar name from the English pronunciation of Ghent, the Flemish industrial city where he was born to Edward's Low Countries' queen Phillipa of Hainault. John was the richest and most elegant and most feared lord on the European scene in his lifetime. He was a patron of scholars and poets. He protected the controversial and heretical theologian John Wycliff, a disgruntled Oxford academic who failed to get tenure, from onslaught by ecclesiastical courts.

Gaunt was the main patron of the poet Geoffrey

Chaucer, who worked as a diplomat representing the English crown on the continent when he was not enjoying the sinecure Lancaster awarded him, chief of collector of customs of the port of London, a sweetheart job. Chaucer's wife was one in Gaunt's large stable of mistresses.

To us the lives of John of Gaunt, the royal duke of Lancaster who bestrode not only England but France and Spain like a colossus for two decades, and his little sister Joan, dead at fifteen on her way to marry a Castilian prince, seem like a study in sharp contrast, indeed, a pathetic one.

But it cannot be overemphasized: to the high nobility of mid-fourteenth-century Europe this facile contrast was not evident. In aristocratic consciousness, John and Joan alike lived out their royal, hyperaristocratic lives, the ones that God and faith had allotted them. Their individual selves weighed equally in the balance of life and history.

The great nobility were not reflective people, even if they were connoisseurs of art, collectors of rare books, and patrons of scholars, poets, and theologians. These latter high cultural pursuits, to which the Lancastrian family actually inclined, were considered in context: just part of a showy, expensive existence. In this life, a prized horse was the same as a patronized poet, elaborate sexual experience the same as none at all, their immensely showy clothes and exquisite dinners just part of the diurnal process—like breathing and sleeping.

This was what they were. They were tactile, existential

personages, not reflecting on long-range or even tomorrow's significations. They were born to immeasurable wealth and great station; they behaved accordingly, in that way no different psychologically from the peasants who numbly pushed plows and existed on porridge. You played the hand you were dealt, the life to which Christ had called you, and then it was over, frequently in childhood or adolescence, almost never after the age of fifty.

The nobility lived these short lives without a sense of irony. Funeral sermons delivered over the coffins by mumbling bishops might indeed expatiate on the shortness and fragility of human life. But the nobility did not act that way, preferring more the visceral contact of the hunting dogs and hawks they loved than the anxiety-ridden, memory-dominated self-consciousness of affluent and well-educated people today.

This life was lived by very few. In 1340, 60 percent of Western Europe's wealth and nearly all its political power were in the hands of some three hundred families of the higher nobility, of which there were about four dozen in England. Their wealth was literally incalculable, since it was never assessed or audited. But the income of each family was at least a billion dollars a year in today's money.

There was plenty of talk among these four dozen families about their own political power. They were the "first estate" or political class of the nation. In England the head of each billionaire high noble family, and often his heir as well, along with some thirty bishops, was summoned by personal invitation of the king to sit in what was coalescing as the House of Lords, Parliament's upper chamber. The House of Lords had both great legislative and legal power (the highest court in England is still the House of Lords; their work is actually done by twenty-five professional lawyers given noninheritable titles for life).

Yet the first estate rarely played an important role in politics, legislation, and law. Contrary to the anachronistic liberal dreams of nineteenth- and twentieth-century historians, the great aristocracy in the fourteenth century did not accomplish much in politics and legislation. Whenever they bestirred themselves to take an active role, after generating a momentary crisis by impeding the royal administrators and drawing up some sonorous oligarchic reform placing the government in their own hands, they very quickly lost interest. The only issue that could truly engage the House of Lords for a few months was the hateful pursuit of some royal favorite, usually gay. That normally ended in violence and the great men then dispersed to their country estates and resumed their well-tilled behavior of feasting, drinking, hunting, and sex.

The impact of the plague on the higher nobility was individual rather than collective. Their cash flow was so huge, their lifestyle so lavish, that they had a significant influence on the economy. Thousands of people in food purveying, building, luxury clothes, household services, horse breeding, as well as weapons manufacture and many more trades and subtrades found much of their market in satisfying the few dozen at the very top of the social ladder.

The pace of life these top nobility set and the luxury goods they cultivated also had the effect of pressing the less affluent nobility and the upper stratum of the middle-class gentry to imitate them as far as more constrained resources allowed. Living on credit became as common among the landed classes as it is in American society today. The Florentine and southern French bankers allowed huge debts to be run up at very high interest rates. The bankers' loans were ultimately safe if given to the highest of the aristocrats because their immense wealth—however lavish their spending—eventually made repayment probable. This was the conventional wisdom with the greatest borrower of them all, King Edward III—until he had to float such huge loans to fund his wars that he finally defaulted, sending a couple of illustrious Florentine banking houses down in flames.

Here then is the peculiar way that the Black Death ultimately affected the awesome Plantagenet family. When Princess Joan was struck down, the dynastic union with Castile was precluded. Even later invasions of Spain, first by the Black Prince and then by John of Gaunt, could not compensate for the lost political advantage of this failed union. But by striking down Henry of Grosmont the plague opened the way for John of Gaunt to inherit the vast dukedom of Lancaster, an event that centrally shaped the next century of English political history and split the Plantagenet family.

In view of the Florentine moneylenders, this was a very good thing. John of Gaunt was able to remain solvent, and set the leading edge in the billionaire lifestyle. He built the biggest house in London, which was burned down by angry peasants in 1381.

CHAPTER FOUR

Lord and Peasants

I N THE YEARS OF THE BLACK DEATH England was still an intensely rural society. Ninety percent of the population lived on the land, engaging in intensive cereal agriculture, sheep and cattle ranching, or both. The country's largest city, London, did not have more than 75,000 people. There were three towns in the 5,000 to 10,000 range—York in the north, Bristol in the west, and Lincoln in the central region. There were perhaps twenty more towns, crossroads on highways and locations for annual commercial fairs or church centers, in the 500 to 5,000 population category. But in these small towns many of the population were still engaged in agriculture in the surrounding countryside.

Nevertheless, England was a wealthy society. Different regions excelled at different kinds of agriculture. North of Cambridge to Liverpool was ranching country, dominated by huge sheep and cattle granges. In this northern region there were four or five times the number of four-footed do-

mestic beasts as human beings. The sheep produced millions
of pounds of high-grade wool per year. It was baled and exported to the industrial textile cities of Flanders (Belgium),
such as Ghent and Ypres, and even further afield to Italy.

At least a third of the raw wool came from the estates of
church corporations, particularly Cistercian monasteries.
They had received unused scrub land from lay patrons in the
twelfth century and had turned these hills into great sheep
ranches. The Cistercian monks became so rich that secular
lords imitated on a grand scale the ranching development
the monks had pioneered. (The Cistercians are still today assiduous businessmen. The Trappist jam and "monk's bread"
we buy in supermarkets today are products of their inheri-
tors, the name a seventeenth-century derivation of the medieval Cistercian Order.)

An elaborate collection, retrieval, baling, and shipping
system carried the wool to southern and eastern ports to be
shipped to the Continent, particularly to Calais in northern
France, a distribution center. The royal government since
1275 imposed an export tax called "the Great Custom" on
this wool, producing in 1340 about 5 percent of the annual
income of the Crown. The government, the landlords, and
merchants represented in Parliament were forever disputing
whether the custom rate on wool should be raised to give the
king more income. This was still a contentious issue in the
seventeenth century.

By 1348, perhaps 10 percent of the raw wool was retained at home, not exported. Entrepreneurs were setting up a native cloth industry, which would expand rapidly in the following century. The cloth was produced not in congregated factories as in modern times, but in the homes of peasants. Agents of the capitalists would go out periodically and give instruments and raw wool to the domestic spinners, collect the cloth they had produced since the last visit, and pay them. Historians have called this prefactory organization the domestic or putting-out system.

Old sheep were slaughtered for their meat after years of giving wool, providing the prized mutton chops for middle-class and working-class tables. (Lamb chops were almost unheard of—the young sheep were too valuable for their wool.) The millions of cattle that grazed on the slopes of northeastern hills were wanted for their leather but especially for their red meat. The Anglo-Saxons who conquered Roman Britain between A.D. 400 and 600, like other Germanic peoples, were intensely carnivorous, devourers of red meat. The more affluent classes ate nothing else, except for occasional exotic fowl, morning, noon, and night.

For centuries red meat had been amply available in the form of venison from deer slain in the forests. But over the centuries the forests had been greatly reduced in size for purposes of agricultural development and peasants' village settlement. By the fourteenth century the unquenched English upper-class appetite for red meat could only be satisfied by the raising and slaughtering of domesticated cattle.

Animals raised under crowded conditions were prone to cattle epidemics, of which the most menacing was anthrax. Some time in the fourteenth century, probably around 1340, a strain of anthrax was communicated to humans just as the origins of HIV/AIDS came from transmittal of infectious disease from chimpanzees to humans in East Africa some time between 1930 and 1950.

Fourteenth-century doctors never identified the emergence of an anthrax epidemic among humans. Because the first stages of bubonic plague and anthrax are identical—flu-like symptoms and high fever—they thought the anthrax attack on human society was the familiar bubonic plague. Some physicians were puzzled that a minority of plague victims never developed the distinguishing buboes, black welts around groins and armpits, that give bubonic plague its name.

They did not draw the conclusion that some of the pestilence's victims were actually succumbing to anthrax. In spite of the crowded conditions today on western American beef-producing ranches, anthrax is now prevented by annual inoculation of vaccines. Otherwise it would likely spread as in the great ranches of northern England and the small pasturages in the south in the fourteenth century.

The central and much of the southwestern part of England was called champion (open field) country—the rich agricultural, cereal-growing heartland that produced at least half the country's wealth. Because of advantages of soil and weather conditions, this farmland was agriculturally among the most productive in the world. Only the Ukraine and parts of western Canada and the U.S.A. are as propitious for growing grain as the champion part of England, about 40 percent of its total land mass.

From 1870 to 1940 short-sighted governmental policies involved the abandonment of much of this intensive agriculture and the English ate bread made from Canadian and American grain. During World War II, because of the effectiveness of the German U-boat blockade until 1943, the British government tried desperately to reverse this forsaking of the land, dispatching "land girls" from the city and even professors from Oxbridge to hastily resurrect farm production, with modest results.

Down in the English southwest in the counties of Devon and Cornwall and in Wales, which the English crown had conquered in the late thirteenth century, the land was mostly too chalky or rocky for agriculture, and the population was thin and impoverished. Coastal villages lived off smuggling. In Wales a primitive coal industry was slowly emerging, because by deforesting their once richly wooded country, the English had begun to experience early signs of a puzzling trend toward a fuel shortage that would become critical by 1500.

In south central England, the heart of champion farmland, however, the century from 1180 to 1280 had been the medieval golden age because of favorable climatic conditions. The climate of the northern hemisphere, including

England, experiences alternating cycles of warming and cooling. A warming trend had set in during the early twelfth century and it reached its height in the century after 1180. It was a time of long, warm summers and moderate winters. There always seemed to be enough rain to make the cereal crops sprout fervently. There were no crop failures or famines.

More rural working-class young people, whose diets were heavily dependent upon cereals, survived to adulthood, and life expectancy was extended. The rural population skyrocketed, tripling in the thirteenth century. By 1280 England's population was approaching six million people, three-quarters crowded into the agricultural south central heartland. Not until the middle of the eighteenth century would England's population again rise to six million (it is sixty million today in a heavily urbanized society).

The downside of good weather and sharply rising population was an unprecedented boom in agricultural real estate. The thirteenth century in England was a time of land hunger. Every arable inch in the rich black earth champion central region was put under the plow. Millions of acres were deforested and settled with peasant villages. The space preserved for grazing cattle and sheep in each village was cut back. Less attractive land, on hillsides or on more chalky soil up to now ignored, fell to efforts at cultivation. Aerial photographs today show the palimpsest of these villages created on marginal land and abandoned after the Black Death had reduced the rural population.

The price of land on the rural market rose very rapidly and steeply in the thirteenth century. The royal courts were flooded with lawsuits over ownership of parcels of land and registration of land-purchase transactions. The new profession of common law attorneys learned their trade in complex land litigation in the county courts.

The landlords pressed the royal government for a parliamentary statute that would unequivocally legalize the selling and buying of land and wipe away hoary restraints on rural capitalism derived from judicial detritus of earlier centuries. Courses on property law in the first year of American law schools today begin with this resulting statute of *Quia Emptores* of the 1290s, which unequivocally established a capitalist free market in the land.

At the same time as land value kept increasing, landlord families got their lawyers to work out elaborate restraining documents preventing their heirs from ever selling off family properties. The noble and gentry landlords wanted to preserve the family estates intact to the end of time. These complex legal instruments, called entails, were still key plot devices in some of Jane Austen's novels, written in the early years of the nineteenth century. In the 1670s a progressive judicial decision stipulated that land could only be entailed for one generation at a time, not perpetually. It was not until 1833 that the vestiges of the late-thirteenth-century entail system were finally abolished by act of Parliament.

In the frenzy of the liberated capitalist land market, the

legal status of serfdom that had been imposed on the peasantry around the beginning of the second Christian millennium became increasingly obsolete, pointless, and actually dysfunctional.

Serfdom had been meant to ensure a steady supply of labor by legally tying generations of men to the land (if your father was a serf, you were one too). Serfdom exists where land is cheap and easily available but peasant labor to work the land is in short supply, as in the sixteenth-century Ukraine or eleventh-century England.

English serfs were not slaves—human chattels as in the Roman Empire and American South. They had legal rights, and the system had heavy costs for their lords. Serfs had a right to strips of arable land of their own to work (after putting in around two-thirds of their time working the lord's personal lands, called his demesnes). The serf villagers had a right to pasturage of a modest number of domesticated animals. They could hunt for boar and rabbits (not deer, which were reserved for the ruling class) in the neighboring forests or haul fish out of a nearby stream to eat on meatless Catholic Fridays and during Lent. They could plant and cultivate vegetable gardens next to their houses. The lord had to provide in each village a mill to grind the peasants' grain for their heavily cereal diet. And not least, the lord of each peasant village had to build a humble church for the peasants and staff it with a more or less literate priest to perform church services.

Serfdom should not be identified with a starving and abused peasantry. There were plenty of fat and prosperous serfs in England in the 1300s, like the kulaks in the pre-Stalinist Soviet Union.

There were also in the thirteenth century ambitious serf families who sought freedom as their legal status. That would make them mobile; they or their sons could leave the ancestral village and set up a household elsewhere. As freemen they could accumulate land from neighboring peasants through purchase and emerge as what around 1400 came to be called yeomen—free peasants. Those yeomen families with vaulting ambition aimed to marry or buy their way into the lower echelon of the gentry class, the landlord class.

After 1180 manumission of serfs became a steadily increasing trend, as with the population boom landlords realized that the old labor shortage had been superseded by a fluid labor supply. Now there were strong peasant backs and calloused hands to hire as laborers on monthly or yearly contracts. The labor market—turning steadily more advantageous to the lords in the century after 1180 as the rural population expanded—meant that the costs and restraints that serfdom imposed on the lord could be rudely jettisoned. Now the landlord could just hire workers in exchange for cash. As the villagers grew older and less capable of hard labor, the lord would not renew their contracts and would turn them loose on the highways, and into the town and forests. The towns restricted immigration. In the forests, if the peasants were not too old they could join criminal groups, such as the one headed by Robin Hood.

Thirteenth-century English society learned how interconnected were freedom and capitalism, what a Cambridge anthropologist has perhaps over-enthusiastically called "the origin of English individualism."

The royal justices as they went on circuits to each county town twice a year "holding the assizes" (as is still done in the outback of western Canada) abetted the process of eliminating serfdom and giving a legal status of freedom to the manorial peasants.

The trend was also in accordance with the policies of the crown. The Plantagenet monarchy was ever searching for ways to improve its tax base. Serfs did not pay taxes directly to the crown; you had to be a freeman to pay rural taxes. The manorial lord was supposed to collect taxes from his serfs when a general tax levy was imposed and then turn this money (along with the lord's own tax payment) over to the crown.

But the lord would try to cheat and pocket at least some of the taxes he had collected from his serfs and not turn it over to the royal tax collectors. It was therefore to the fiscal advantage of the crown to multiply peasant freedom, thereby bringing the peasant family directly under the king's tax levies.

The royal justices after 1180 found a convenient loophole in the law, which stated that only a freeman could be a litigant in a civil property action in the royal courts. The judges proceeded to allow just about any peasant who wanted to do so (and pay court costs) to appear before them in their county circuit court without inquiring into the litigants' free or servile status.

Thereby the royal justices automatically granted free status to any peasant who was a plaintiff or defendant in a lawsuit in the county court. This happened on a large scale in the century after 1180. It also by itself dribbled more money into the Plantagenet treasury, because litigants had to pay court costs to have their cases decided in the county courts before the royal justices and the loser even in a civil (noncriminal) action had to pay a small fine to the crown.

By 1280 at least half the peasants in the agricultural heartland of south central England had gained their legal freedom, sometimes by obtaining their manumission from lords, but also by arbitrary action of the royal justices.

In the next hundred years the proportion of peasants still in the serf category shrank with each passing decade. Serfdom and capital land market are an awkward fit: The former is an effect of a status society, the latter of a money economy. The official opening up of the land market in the 1290s by parliamentary legislation accelerated the eventual disappearance of serfdom, developed for a very different society.

In the years that followed, the status of the rural peasantry was a hodgepodge of conflicting judicial mechanisms characteristic of a rural society in transition to a modern real estate market. There were some affluent serfs and plenty of impoverished free peasantry. Yet some of the free peasantry had gotten into the booming land market, had bought up their less enterprising neighbors' lands or purchased more land from a cash-starved landlord (whose attorney had circumvented an entail) and were in transition to yeoman status and ultimately, by 1500, entry into the lower ranks of the gentry landlords.

Climatic cycling continued to drive social and economic change. Around 1280 the warming trend began to run down. A new weather cycle unevenly but visibly intruded into rural England. Summers became cooler and shorter, the long autumns ideal for bringing in the lush crops truncated. Winters became longer and more harsh. The cooler period was to last until the late fifteenth century, when it would be followed by another warm century and then the "little ice age" of the seventeenth century, when people actually skated on the frozen Thames—not something you would want to try today.

In the summers of 1316 and 1317 rural disaster struck. The sun did not shine. There were widespread crop failures. There was famine and death from hunger. These terrible years had a special cause. Huge volcanic eruptions in Indonesia threw continent-sized clouds of ashes into the atmosphere and by 1316 this cloud of unbeing had reached England. Even when the sun shone again and the famine subsided, there were adverse weather conditions—too much rain—for good cereal harvests. The price of grain escalated.

The stomachs of the peasants were no longer full.

Even aristocratic temperament became meaner and angrier. In 1327 a palace coup engineered by Edward II's French queen, Isabella, and her aristocratic lover, with acquiescence on the part of the short-tempered and unhappy nobility, resulted in the king's forced abdication in favor of his very young son Edward III.

The dispossessed King Edward II was killed by a red-hot iron poker shoved up his anus. This savagery partly reflected hostility on the part of the Church and other opinion-makers to the king's homosexuality and his favoritism toward his young French male lover, but it also reflected the general malaise, anger, and pessimism of the new age of global cooling.

It may be speculated that the Great Famine and global cooling of the early fourteenth century and the deterioration in the diet of the common people that resulted had some adverse impact on public health. Undernourished bodies were more easily prey to the Black Death. Did a decrease in the standard of living in northern Europe, including England, open the way for a pandemic of infectious disease? The current state of biomedical history does not allow more than speculation on this matter.

A third of the best arable land in England in 1346 was owned by church officials—bishops and abbots—or by ecclesiastical corporations, the chapters of cathedral priests, called canons, or by monastic communities. Often—although not without noisy litigation—the bishop or abbot directly administered these corporately owned lands as well as the estates directly conjoined to his office.

Most of this land had been granted to the church officials and corporations by lay nobility and gentry in the period between 1000 and 1200 when land was still cheap in England. These grants, set down in elaborate formal land deeds called charters (literally "documents"), were not mere acts of charity like a billionaire endowing a university laboratory or dormitory today. The grant to the church meant a specific spiritual service of great value in the afterlife.

The bishops, abbots, and cathedral canons or monks promised to pray perpetually and even daily for the soul of their benefactor and designated members of his family until the end of time, speeding them through the transitional state of Purgatory where those souls were doing time in recompense for earthly sins, until they were washed clean and ascended to heaven. (Almost no one in the optimistic Middle Ages was consigned to hell in the afterlife, no matter how many or heinous their crimes, except for heretics, witches, and Jews.)

Some of these lands granted by charter to church officials and corporations were already well-developed, heavily populated estates, which heirs would agonize over losing. As the old lord lay on his deathbed no one minded visits by his mistresses and whores. It was the whispered conferences with a priest that concerned heirs, because this could lead to a deathbed alienation of a share of the about-to-be-inherited patrimony. The courts informally decided that no lord should alienate to the church more than 10 percent of his entailed estates. But this was more than enough to enrich ecclesiastics.

With the population boom and real estate frenzy of the thirteenth century, the value of these landed gifts, their charters enforced by the royal courts as well as by ecclesiastical curses against infringement, immensely increased in value. Real estate development and inflation was as well known a phenomenon in the Middle Ages as the miracles of the Virgin Mary. What in 1000 could have been marginal, thinly populated land or vacant scrub land or forest was by 1280 thickly settled prime arable estate or in the north richly green cattle and sheep ranches.

Bishops had for a thousand years been men of business, public officials, usually of noble family, experienced in managing income and skillful at getting ever more of it. How else were the great Gothic-style cathedrals of the thirteenth century in episcopal cities paid for? But abbots in the mists of earlier times had been elected from among their communities for their promise as spiritual leaders or even fame as clergymen and scholars.

By the late twelfth century in England this kind of monk was rarely chosen as abbot. The typical abbot had entered the monastery at any time between age six (as an "oblate" given up to God by his family) and twenty. He filled a succession of administrative offices within the billion-dollar corporation that a large and well-endowed monastery represented.

When an abbot died, the monks gathered in their chapel, prayed for divine enlightenment, and chose from among the community an experienced administrator who had already proven his mettle in property management, accounting, and the perpetual land litigation that a great monastery was involved in. If the king approved of the abbot chosen by the monastic community—and normally he did, since the nominated abbot was a man of the world the crown could do business with, a person of conservative temperament and realistic attitude—the abbot-elect entered into the office for life.

By monastic rules drawn up around A.D. 550, the abbot had complete authority over both the spiritual and physical sides of monastic life, over the bodies and souls of the monastic brethren. Now he no longer dined with the monastic community at its two meals a day, one around 11:00 A.M. and the other at 5:00 P.M. Only rarely and then on festive and ceremonial occasions did he dine with the brothers. He inhabited a house separate from the monks' dormitories and workhouses. He entertained princes, lords, and gentry in his private dining room and put them up overnight in well-appointed bedrooms.

The abbot surrounded himself with a couple of secretaries and a dozen administrative officials, drawn normally from the monastic brethren. He hardly ever made a pilgrimage to a religious shrine, but he was frequently in the saddle accompanied by a bodyguard inspecting work on far-flung estates of the abbey. He retained a secular lawyer, frequently conferred with him, and appeared frequently in the county court or even one of the central courts of London to give testimony in major lawsuits affecting the abbey's property. He engaged in year-long wrangles with the king's treasury officials over minutiae in the abbey's tax returns.

The abbot was forever on the lookout for expanding the monastery's landholdings, either through major gifts from laymen (not so easy now to come by as in the good old days of cheap land and before heirs had learned to hire their own watchdog attorneys) or purchase.

We must think of the fourteenth-century abbot as something like the president of a leading American university today. The latter usually has an academic background but he is not selected for those skills. He became a corporate chief executive officer, a man of business, a bigtime capitalist manager. So it was with a late medieval abbot. Such an official had to be forever attentive to changes in the social environment that could affect the abbey's wealth and social status—the ambitions and imbecilities of kings, dramatic changes in climate, and the impact of pandemics on the peasant workforce. These vicissitudes were central to the abbot's career.

Although he had absolute rule over his community for life, the abbot had to keep the monks happy, which meant prosperous, secure, and above all, well-fed. The kitchen records of Westminster Abbey in London from the mid–fifteenth century have been discovered and analyzed by Barbara Harvey in *Living and Dying in the Middle Ages* (1993), a sensational book that caused a lot of embarrassment to pious medievalists.

The results show that each of the sixty monks consumed two pounds of red meat a day, plus countless morsels of fowl and fish, washed down with claret wine or fresh ale in unlimited quantities, buttressed by an array of sugary desserts. Not all of the thousand abbeys in England could afford this lifestyle, but probably half could and assiduously cultivated it.

Records of fourteenth-century gentry households indicate that the food consumption per capita in secular upper-middle-class landed families was about the same as in monastic communities. But weren't the monks supposed to be ascetics? Some people thought so, especially wise-guy poets like Geoffrey Chaucer. But in fact what made you a monk (or nun) was living a regular, disciplined life in a cloistered community, not abstaining from a heavy gentry diet. The abbot, whatever else he did, had to make sure that this high calorie consumption was not diminished. That alone was a challenge to the abbatial executives.

But force-feeding the fat monks was relatively trivial compared to the problems raised by the unprecedented mortality of the Black Death in the late 1340s. The sudden striking down of perhaps a quarter of the monastic community meant an unusual number of vacancies among the brethren to be judiciously filled. More important, 40 percent of the labor supply in the manorial villages on the great estates owned by a large monastery disappeared.

Having gotten rid of the plague-infested bodies, usually in mass graves with victims stacked like cordwood (or as a contemporary Italian writer remarked, like lasagna), five cadavers deep, the abbot had to sort out a bewildering array of problems arising from the vacant tenements and the threat of a sudden shift from a glut to a dearth of agricultural wage laborers.

Close study by economic historians has revealed that the ecclesiastical executives as a whole did better than might be expected—for a while. The most severe impact of the population collapse was postponed in many places for a generation until the 1370s, because in 1350 there were still so many landless peasants ready to take over vacant rent-paying farms. Yet for an abbey that already was mismanaged and heavily indebted, the Black Death represented a crisis that imposed an additional burden on the corporation's bottom line and high anxiety for executive abbots. These were businessmen under pressure.

On the estates of the abbot of Halesowen in Worcestershire in the rich farmland of central England, the impact on and the response to the pandemic by a very rich abbot and hundreds of peasants who worked for him and paid rent to him can be scrutinized, thanks to Zvi Razi's research. It is the kind of business case study favored today by American business schools.

Faced with an unprecedented physical disaster that threatens the fiscal viability of the corporation and its way of doing business, how does a skillful executive officer respond? And—less interesting to the business schools—how are the lives of the corporation's workers affected? Halesowen gives us medieval answers.

The great manorial estate of Halesowen with its thousands of peasant families obligated to the abbot in one way or another was located in what is today the suburbs of the modern metropolis of Birmingham. This industrial city did not exist in the fourteenth century except as a small village. The big city for Halesowen was the cathedral center of Worcester, where the judgments and opinions of bishops were meaningful although not necessarily authoritative (such were the complexities of the church's canon law) for the abbot executive of Halesowen.

The abbot of Halesowen when the plague struck in 1349 was Thomas of Birmingham, a member of an old country family after whom the modern city is named. Thomas of Birmingham had been abbot since 1331. He survived the plague and remained abbot until he died of natural causes in 1369. Given the short life span of fourteenth-century men—although the well-fed monks shielded from violence usually outlived the forty-year life expectancy for laymen—Thomas's long tenure as abbot was unusual. Imagine someone today as chief executive officer of a midsize corporation for thirty-eight years—uncommon. Thomas of Birmingham's successor, William Bromsgrove, survived old Thomas very briefly, being cut down in a recurrence of the Black Death in 1369. Another election had to be held, making two in one year, an unsettling phenomenon for the monks.

None of these abbots took much part in politics or local society beyond the abbey. They were totally absorbed in the religious and especially economic life of the monastery. The Birminghams had stayed out of politics but William, Abbot Thomas's father, was a member of the entourage of a high-level gentry person, John of Sutton, who hazardously dabbled in national politics in the late 1320s. When John of Sutton got in trouble with some royal court politics, William of Birmingham acted as guardian of the Sutton estates to keep away marauders and enemies.

Before the plague decimated the peasant population on Halesowen's lands, Abbot Thomas had faced some stiff challenges that threatened the abbey's income. On the positive side he had been able to obtain three modest grants of land from local gentry. The largest of these came from an heiress on the conventional condition that candles be lit in the monastic chapel to accompany the prayers sung on behalf of her soul. The abbey was also to distribute twenty shillings (about one thousand dollars) annually to the poor in her memory.

Abbot Thomas successfully negotiated with the royal government and the bishop of Worcester to be allowed to appropriate these gifted lands into Halesowen's core estates.

But in his petition to the bishop, Abbot Thomas referred to recent events that had harmed the abbey's income. A serious fire had destroyed houses and shops in the town of Halesowen itself. This hurt because a substantial part of the abbey's income came from commercial activity—taxing the borough merchants and conducting a weekly market in the village square and a four-day fair twice a year. Each fair was conducted to commemorate one of the two saints to whom the abbey was dedicated, St. Barbara and St. Kelem.

Kelem was a real person, an eight-year-old Anglo-Saxon prince, murdered by his greedy sister and her lover in the eighth century. St. Barbara was an entirely mythical saint of Middle Eastern provenance supposedly martyred by a Roman emperor in A.D. 303. She was thought to offer protection from sudden death, usually as a result of lightning or cannonballs.

Abbot Thomas complained to the bishop of Worcester that St. Barbara had lost her drawing power and this affected attraction of gift-bearing pilgrims and business at the fair dedicated to her. Possibly the "coldness" to St. Barbara that Thomas said had developed among the populace was due to her failure to help in the great famine in the second decade of the century, which was followed by bad weather and an outbreak of murrain. She certainly did not help in the Black Death.

The abbot also complained to the bishop that the abbey was getting the wrong kind of visitors. The costs of hospitality to travelers and wayfarers (monasteries were the prime motels of the Middle Ages) were a heavy burden. The latter complaint was likely a trope, a literary formality. But it does seem that the monastery was suffering diminished income.

Halesowen kept precious (alleged) relics of St. Barbara and St. Kelem encased in silver and gilt reliquaries (saints' relic boxes). Not only were saints' relics intended to draw paying pilgrims to the abbey, but they were trotted out publicly in critical times. And there could not be anything more critical than the Black Death of 1349, when 40 percent of the monastery's tenants died, threatening to devastate the abbey's income and deprive the monks of their accustomed daily feasts. Processions of saints' relics through the streets and along the roads were the most immediate communal responses to crisis (they still are in some Mediterranean countries and Latin America).

As word of the biomedical catastrophe spread and his agents reported uniform mortality from distant villages belonging to the abbey, Abbot Thomas sat in his chapel in the gloom of a late afternoon and wondered why St. Barbara and St. Kelem had not interceded for Halesowen. He did not know that St. Barbara was a fake, nor would he have believed it if someone had told him so.

Religious expression not being effective against the plague, Abbot Thomas had to exercise his managerial skills to deal with its impact on the abbey's income.

How a great estate worked by thousands of peasants responded to the Black Death reflected prevailing conditions before the plague. Peasant holdings that remained deserted after 1349 reflected the quality of the land. If the peasant landholding was marginal in the first place, having been brought under the plow during the population boom and real estate inflation of the late thirteenth century, now that 40 percent of the village population had died in the plague, there would be no rush to take up these hardscrabble marginal lands.

They would remain vacant and become the "lost villages" of medieval England, disappearing from written records and only rediscovered in the twentieth century by air photographs.

Abbot Thomas was fortunate compared to some other lords in that Halesowen lands were mostly of high quality. This meant that there were surplus laborers in the peasant population eager to step into the breach and take up these vacant lands, which thereby would continue to generate revenue to the abbot's treasury. Zvi Razi concluded that the great majority—82 percent—of the vacant farms on the abbey's estates were quickly taken up by new farmers drawn from the surplus population or were added to existing family holdings.

There is no evidence of peasant flight from Halesowen to other estates. On the contrary, if some Halesowen land vacancies were somewhat difficult to fill, immigration from elsewhere boosted the rent rolls over time. Abbot Thomas benefited from an important concession that his predecessor as head of the abbey had made in 1327 after almost a half century of litigation and agitation by Halesowen's peasants.

They wanted to remove the last vestiges of serfdom, abolishing the mandatory servile labor on the lord's demesne or personal estates, and commute these services into money rents. Not without some turmoil and punishment of one of the peasant ringleaders, this concession had been granted. Now in 1349 Abbot Thomas benefited from this legal and economic change, because it made peasant holdings on the abbey's estates along with the intrinsic high quality of the vacant farms more attractive than other great manors where bothersome incidents of serf labor lingered.

That 18 percent of Halesowen's tenant farms remained vacant was not very burdensome to the abbot as, like most other great monasteries, Halesowen did not, in any case, rent out all its lands for peasants' rents. It kept some land as demesne, for direct cultivation under the abbey's management. These lands contributed to the food that filled the monks' stomachs in their high-consumption lifestyle. Surplus agricultural products from the demesne were sold at local markets and fairs—in 1369 this surplus returned almost eighty-five pounds to the abbey's treasury (three hundred thousand dollars in today's money). Most of it was used to repair farm buildings on the monks' estates, keeping up those barns and storage sheds that today have become desirable country homes colorfully advertised in the real estate sections of the London Sunday newspapers.

The iron law of supply and demand, however, worked against Halesowen, as it did for many other landlords in the three decades after the Black Death. The economic historian John Hatcher showed the squeeze in labor costs for the land-lords did not come immediately after the catastrophe of 1349, but a generation later, in the 1370s. By then the sur-plus laborers who had eagerly and gratefully taken up rent-paying farms were all gone.

The population level did not recover from the plague, did not resume the skyrocketing demographic curve of the late thirteenth century. Population growth was halted by the great famine of the second decade of the fourteenth century, and population level was calamitously driven downwards in the late 1340s from infectious disease.

The climatic and biomedical shocks of the fourteenth century induced caution in the peasant population just as the Great Depression affected the Great Generation of penny-pinching Americans who went heroically to fight World War II.

The more conservative behavior pattern, changed from the boom days of the thirteenth century, meant that by 1370 marriages among the common people, perhaps even the middle-class gentry, were occurring significantly later or not at all. A historical sociologist has estimated that in mid-fif-teenth-century England one-quarter of the population never married. Late marriages or long periods of bachelorhood and spinsterhood between marriages were the main social factors in the fifteenth century that kept the population from bouncing back.

Modern economists regard a 5 percent unemployment rate as the lowest a society can experience before labor shortage, workers' discontentment, and inflation in wages occur. Another crunch came from the fall in grain prices in the 1370s as the population leveled off at 40 percent below its peak in the early fourteenth century, reducing demand for foodstuffs on the market.

By the late 1370s Abbot Thomas's successor was caught in a price squeeze—declining grain prices and push of labor for higher wages and benefits at a time of virtual full employment. Having gained their release from serf labor dues on the lord's demesne in 1327, Halesowen peasants were now resisting the customary "boon labor" of haymaking and harvesting that were outside servile status, but were ancient community services that freemen as well as the old serfs engaged in. There is a charming picture of boon service harvesting in Thomas Hardy's novel of rural life in the 1860s, *Far from the Madding Crowd,* and the film made from it.

England at the end of the 1370s was slipping into a revolutionary situation—landlords under pressure of price squeeze resisting demands for higher wages and lower rents and elimination of all communal labor, and turning to Parliament for legislation to withstand the tide of working-class demands. The peasants were trying to improve their position in a labor market favorable to themselves. Radical clerics recently graduated from Oxford moved about the countryside sharpening the class consciousness of the peasants. The result was the greatest medieval working-class rebellion, the Peasants' Revolt of 1381. It inflamed much of the eastern third of the country, came close to bringing down the royal government, and led to the killing of both lay and ecclesiastical lords by peasant crowds and the burning of manorial records that testified to the peasants' obligations.

Halesowen was too far westward in central England to become immersed in the Peasants' Revolt. But echoes of the great uprising and the preaching of Oxford radicals reverberated even there. The peasants in one large Halesowen village unilaterally declared their complete freedom.

In the aftermath of the uprising of 1381 in eastern England, special judicial commissions were dispatched to restore law and order. One such commission reached Halesowen and consigned the leader of the democratic-minded peasants to prison, where he died.

The loosening of the bonds and bounds of rural society caused by the Black Death and resulting in the Peasants' Revolt of 1381 could have led to a working-class takeover of the government and a socialist state. The peasants in 1381, building on the rumbles spreading through all of rural society, including Halesowen, had that possibility within their grasp. The royal government cowered fearfully in the Tower of London while a crowd of many thousands of militant peasants from eastern England gathered in a field in the

London suburbs. But the peasants were naïve and the disgruntled graduate students from Oxford who had helped to articulate the peasants' grievances and demands into a vision of a Christian commonwealth were too bookish and inexperienced to be capable of directing the rebellion toward a Leninist or Maoist denouement.

The young King Richard II came riding out to meet the peasants. He assured them that he loved them and if they would go home their demands would be met and justice fulfilled. The most militant of the peasant leaders was struck down by a Plantagenet courtier accompanying the pallid young monarch. The peasants dispersed and the power of the government, using the instrument of class-biased common law, came down on them hard and hung most of those identified as proletarian leaders.

The main social consequence of the Black Death was not the advancement of a workers' protocommunist paradise but further progress along the road to class polarization in an early capitalist economy. The gap between rich and poor in each village widened. The wealthiest peasants took advantage of the social dislocations caused by the plague and the poorer peasants sank further into dependency and misery. Class polarization, capital accumulation, social mobility into the yeoman class: These were the tangible outcomes of the Black Death in Halesowen as elsewhere.

A handful of Halesowen peasants are more than statistics. William Thedrich served as juryman thirty-five times in his life, reflecting his wealth and influence on the manor and among the villagers. He was a legalistic man, suing seven villagers for debt. He was also violent and was fined in court eight times for assault. In fourteenth-century England—as in the late-nineteenth-century United States—litigation and violence were alternative instruments of capital accumulation and social mobility.

William's son Thomas moved swiftly to take advantage of the plague. He was married to a daughter of another rich peasant family, the Thomkyns. When the three top men in the Thomkyns' clan died of the plague, Thomas Thedrich was in a position to claim their lands. He made arrangements with Abbot Thomas, paying him off so that he could lease half of the Thomkyns' lands. He ended up as guardian of the young sons of the Thomkyns family, giving him additional instruments for land accumulation, with the bribed abbot's compliance.

The Moulawes were another rich peasant family who benefited from the plague. They were known for the large size of their cattle herd at a time when many poorer peasant families had none at all. John Moulawe succeeded his father William, who died in the plague. He continued the family's skillful management and bold real estate transactions. John Moulawe married the heiress of another rich peasant who bought him fifteen acres as her dowry. Then in 1355 he bought out his sister-in-law's share and now had enough land to form the basis of a yeoman farm.

The manorial records show how difficult it was for those who started out with little to accumulate more. The court records list those whose holdings were so small that when they died in the plague their heirs were not required to pay heriot, a nominal inheritance tax, because they were deemed too poor. Other entries in the court record illuminate the pitiful lives of servant girls whose desperation led them to steal from their employers and who were expelled from the village, to die as beggars on the highways.

After expulsion from the village, these peasant servant girls' best prospect was to get to London and work as prostitutes. Most of them would not have made it. They died dressed in rags from hunger and disease along the way. They would beg for food and shelter at the doors of cathedrals and the gates of monasteries and nunneries. Sometimes they would get a handout. Most of the time they would be turned away.

Late medieval England was not a welfare society. That did not begin to happen until the application of the Elizabethan poor laws in the 1580s which, however, treated the able-bodied poor as prisoners in workhouses and gave the rest starvation-level aid. By a broader and more humane definition the English welfare state did not begin until the Labor government of 1945–51, and Margaret Thatcher would have loved late-fourteenth-century and fifteenth-century England.

It is known from legal and economic records what peasants as a group did. Getting at thoughts and sensibilities of the peasants is another matter. There is no surviving peasant writing. Perhaps as much as 5 percent of the peasant population in 1350 was minimally literate. Possibly they wrote down their personal reactions to the great pestilence but thus far these writings have not yet been discovered. Even what we know about the greatest working-class event of the fourteenth century, the Peasants' Revolt of 1381, comes in large part from a graphic and detailed account written by a London courtier or cleric, possibly a merchant, from a class level way above peasantry.

Historians have argued that you can penetrate the consciousness of the fourteenth-century peasantry by examining the motifs in paintings and sculpture found in churches of the times. Art is held to have been the poor man's Bible: It was put there, it is assumed, so that the illiterate peasantry could visually look upon the essentials of Christianity.

There is some truth to this. But the decisions on the motif to be presented to the populace were made by the bishops, abbots, and cathedral clergy and the monks who hired the painters and sculptors (unless, as was the case perhaps a third of the time, the ecclesiastics were doing the artwork themselves). The motifs, or what art historians call the iconology, had to be meaningful and inspiring if the ideas were to be communicated to the peasants. But it was still an ecclesiastical thought projection. It did not come from the peasants.

From the thirteenth century on Mother Mary and infant

Jesus became steadily more central in church culture. The mature Jesus was now a young man suffering on the Cross, not the majestic Emperor-Christ of earlier centuries. Presumably this feminization and personalization of Christ struck home with the peasants, but we cannot be sure. A late-twelfth-century writer shocks us by saying there were a lot of skeptics and unbelievers running around. But even if we are sure that this art faithfully represents peasant consciousness, it was only part of their lives and thought, perhaps a small part.

The peasants were in 1350 not even required to attend church and take the sacrament of the Mass more than once a year. When they did, it would nearly always be at a humble parish church with little or no artwork, not at an imposing cathedral or abbey church.

In the fourteenth century Franciscan friars moved about the countryside delivering sermons in English to peasants, preaching in front of churches, at crossroads, or in market squares. Some of these sermons were written down and have survived. Since the Franciscans were engaged in direct communications with peasants, the latter's mentality can be presumed to be reflected in these vernacular sermons.

It was a world of struggling to get food, and a fearful world in which the forces of nature were a constant threat to bring about devastation and death. Jesus and Mary were beseeched to help and sometimes they did. Violence, drunkenness, and physical accidents were prevalent.

Once in a while in the court rolls, especially in criminal cases, the peasant's personal response to an indictment is heard. He is choking with fear; he is begging the court to show mercy and not hang him or her, or in the case of some women not to burn her as a witch. There is constant dispute—over boundary markers between plots of land and over rights of pasturage. The structure of Catholic theology seems as lacking in the legal rolls as the courtly intonations of Chaucer's romantic poetry.

The peasants in the judicial records have vague identity. Most have no set surnames. They are often called by the name of the village or manor they came from. A surprising number of them are known by nicknames: "Hugh Hop over Humber." Yet this vagueness in nomenclature was still true of many Jewish immigrants from Eastern Europe entering the United States at Ellis Island around 1900: The immigration officials gave them surnames, just as the judicial officials of fourteenth-century England did. Clear identity by surnames and given names is a mark of modern society and the bureaucratic state. These rural workers had not yet arrived.

There is one work of literature written a couple of decades or so after the Black Death—*Piers Plowman*—that helps us understand the peasant world. This longwinded but at times gripping religious epic was written by a London clergyman under the name of William Langland, which may have been a pseudonym. It is a good hunch, but not more than that, that Langland had previously served in a rural parish and had come to know the peasants well. He may not have come from peasant stock himself, as were the majority of parish priests, but from the incipient yeoman class, which explains his literacy. That he wrote in English rather than the upper-class languages of Latin or French does not indicate Langland's background. Geoffrey Chaucer, his contemporary, wrote his most famous work in English and he was a courtier and government official.

What *Piers Plowman* tells us about the behavior and mentality of the English peasantry shortly after the Black Death is what you would find among the farmers in very rural areas today. There was a sharp dividing line between the landholding wealthier peasants, who experienced a cyclic boom and bust in their lifetime—sometimes eating well on something approaching a gentry diet and at other times struggling to fill their stomachs with just about any cereal or scraps of cheap meat they could find.

The pressures of farming life produced a lot of tension within families. There was wife abuse and there was "a wicked wife who will not be corrected; her husband flees from her in fear of her tongue." Many times growing up in farm country in Manitoba I noticed shriveled-up, bitter women like that—and they had a lot to be bitter about.

There are the "poor folk in cottages" who have no land and who exist perilously as day laborers and seasoners. They have "no coin but their craft to clothe and keep them." What money they make is spent mostly on "house-hire" (rent), and they exist on a "mess of porridge," happily once in a while supplemented by scraps of cold meat and cold fish. These "afflicted churls" have too many children. "Crippled with hunger and with thirst, they keep up appearances and are abashed for to beg" (translated by Terence Tiller).

There may be further insight into the consciousness of the wealthier and more literate peasants in a thirty-page English poem, *Pearl*, from the 1370s or 1380s. *Pearl* was first edited and published in the 1920s by the Oxford don J. R. R. Tolkien, later author of *Lord of the Rings*. *Pearl* is an elegy, a dirge, for a dead girl written by a country gentleman within the English midlands a hundred miles north of Oxford, about three decades after the Black Death.

Pearl is a pious poem shaped by traditional theology but expresses in a highly personal, emotional tone a perfect balance between form and content. Very near the end of the poem (here quoted in the splendid verse translation by Marie Borroff) there is a glimpse of another theme: anxiety, confusion, and disorientation:

> *Drawn heavenward by divine accord*
> *I had seen and heard more mysteries yet;*
> *But always men would have and hoard*
> *And again the more, the more they get.*
> *So banished I was, by cares beset*
> *From realms eternal untimely sent;*

How madly, Lord they strive and fret
Whose acts accord not with your content!

Perhaps the little angelic girl who is the subject of this poem, hypostatized by the poet into a precious pearl and seen in a dream—common poetic motifs of the time—died in the Black Death.

❦

But in this twilight world between medieval feudalism and early modern capitalism, the law of hierarchy still prevailed. There was restlessness and even occasional rebellion among the peasants, but the only comprehensible model for social relations was still that of lordship and hierarchy. The way of thinking was still vertical, still from the top down, even if some degree of individualism and a liberal sense of community was starting to creep in.

The church authorities who controlled education and social theorizing still stressed the pragmatic value of the natural law of hierarchy. Tell us who your lord is, and we can then tell you what your social and political identity is—this was the medieval tradition that still prevailed.

The clarity and stability that the medieval idea of lordship and practice of hierarchy represented had strong functional values. You knew your place and who your lord was; if he could be demanding or even cruel, he was ultimately your rock and redeemer, your comforting protector, sometimes gracefully inspired by conscience, but always driven by shrewd consideration of your economic value to himself.

If the lords were struck dumb and made immobile by a great physical catastrophe such as the Great Plague, then times were really bad for peasants, bitingly dismal for the medieval world's ordinary people. The electrical force of lordship and hierarchy was momentarily switched off, leaving a society given to anxiety, grief, and confusion. The peasants living on the estate of the abbot of Halesowen felt this displacement in their bones.

CHAPTER FIVE

Death Comes to the Archbishop

❧❦❧

ON AUGUST 19, 1349, a fifty-nine-year-old clergyman and former Oxford University academic, Thomas Bradwardine, landed at the port of Dover on the English Channel, having crossed over from France. Before sailing for Dover, Bradwardine had traversed a great part of France by horse and foot from the papal court at Avignon on the Rhone River.

Avignon then officially lay just outside the borders of the French kingdom, in the ramshackle territory of the German emperor Charles IV of Luxembourg. But in fact it was a city under French control. Here since the first decade of the century a succession of French popes had lived in exile from Rome, ostensibly because Rome was ridden with civil strife and organized crime.

The Avignon papacy had emerged from the French

monarchy under Philip the Fair, having gotten rid of a trouble-some Italian pope in 1303. Boniface VIII had a heart attack af-ter being taken into captivity at his summer residence in the hills near Rome by three hundred French soldiers commis-sioned to bring the pope to Paris for trial as a heretic. Boniface, in a moment of exasperation during his long and noisy quarrel with the French king over dividing the tax income from the French church, was reputed to have remarked that he would rather be a dog than a Frenchman, and dogs had no souls, so the logic, if not the motive, was clear.

After Boniface's death the French government got the majority of the cardinals, about 40 percent of them French-men, to elect the archbishop of Bordeaux as pope. Although Bordeaux was in English-controlled Gascony, the arch-bishop was attuned to the politics of the French government in Paris. The new pope never made it to Rome. Instead he and his successors (until the second decade of the fifteenth century) established themselves in the pleasant river town of Avignon on the Rhone.

Today Avignon is known for the ruins of its medieval stone bridge over the river (and a folksong that commemo-rates it) and for the summer rock concerts and theater there, as well as for vineyards planted by the popes, which still pro-duce the finest and most expensive red Southern Rhone wine, Châteauneuf-du-Pape.

The huge ugly palace that the popes built in two stages, in the 1320s and 1360s, is still there in good shape and is now a state-of-the-art museum specializing in Picasso's later paintings. Inside, the building is sparse and austere—all the medieval artwork and furnishings were removed in the nine-teenth century when the anticlerical Third Republic used the Palace of the Popes as a stable for army horses. To anti-Catholics horse piss and defecation on ecclesiastical grounds is the ultimate, most satisfying insult. Oliver Cromwell, the Puritan dictator of England, did the same thing to the An-glican churches during the English Civil War of the 1640s.

What is now the main exhibition hall at Avignon's Palace of the Popes was in the 1340s the pope's banquet hall at which fresh Rhone wine washed down exquisite and lengthy feasts. Perhaps the most impressive sight at Avignon today is a huge kitchen in a separate building from the main palace that any three-star Michelin chef would still find commodi-ous, if a little awkward, in its use of massive open-hearth wood-fired stone stoves.

Bradwardine had been in Avignon to get the blessing of Pope Clement VI for his consecration as archbishop of Can-terbury, the top position in the English church. Bradwardine was on the Continent in his frequent role as royal diplomat when King Edward III approved his election to the Arch-bishopric of Canterbury, an immensely important position politically as well as vastly rich and potentially spiritually in-fluential, made by the chapter of the Canterbury cathedral monks. Bradwardine went immediately to Avignon to get the papal consecration. He got it, but not without some sarcastic joshing from the pope and the resident cardinals, mostly Frenchmen by this time.

This was because only four months earlier Edward III had sought papal approval for a different candidate, his chancellor John Offord, even though the Canterbury canons had elected Bradwardine. In doing so the Canterbury clerics had probably acted to please the king, because Bradwardine was Edward's personal confessor and a well-used royal diplomat.

The previous archbishop of Canterbury, John Stratford, 1333–48 (who probably died of natural causes, not the plague), had occasionally been in a state of political tension with the royal government, as often happened with the Canterbury prelates. By picking Bradwardine, the king's confessor, the monks in 1349 thought they would please Edward. They may have done so, but Edward surprisingly declined Bradwardine's nomination—as he had a legal right to—and sought approval from the pope in Avignon for his chancellor, the head of the royal administration, John Offord.

There was plenty of legal precedent for Edward's unilateral action, the most famous being Henry II Plantagenet's appointment of his chancellor and drinking and womanizing companion Thomas Becket as archbishop of Canterbury, a rash act that turned out badly for all concerned, especially for Becket, who was cut down in 1170 by the swords of four royal courtiers as he stood in front of the altar of his cathedral shouting "pimp" at the leading assassin.

Edward's preference for his chancellor Offord—only nominally a cleric—over his confessor Bradwardine, an esteemed church thinker and academic, was peculiar. England was in the midst of the great war with France and the kingdom needed strong spiritual leadership to buttress royal propaganda on behalf of heavy taxation and military service. Offord, aside from having no spiritual reputation, was already sick and paralytic at the time of the king's nomination of him for appointment to Avignon.

Possibly Edward did not want to lose his confessor, who had become a personal friend and courtier as well as trusted diplomat. Or perhaps the king feared that the learned, popular, and ambitious Bradwardine would turn into another Becket. Regardless, the plague intervened and Offord died of the Black Death on May 2, 1349, his decaying body unable to withstand the pestilence.

Thus there was joshing and a bit of irritation in Avignon at the king's seeming high-handedness when Bradwardine, a prominent European figure, turned up at the Palace of the Popes to get the papal consecration. Clement VI remarked that if Edward asked him to appoint a jackass as archbishop of Canterbury, he would have to agree to it. (Yes, he would, as a matter of fact.)

Of course the pope consecrated Bradwardine, an unusually qualified archepiscopal appointment in every respect. But at a banquet in the great papal feasting hall, as a platoon of puffing servants rushed huge platters of delicacies through the ten yards out-of-doors from the gargantuan kitchen, one of the cardinals, a relative of Clement VI, arranged for a clown seated on a jackass to enter the hall and present a petition that he should be made archbishop of Canterbury. The austerely intellectual Bradwardine probably feigned a smile through gritted teeth—just the sort of nonsense that he might have expected at Clement VI's flamboyant court.

After Bradwardine landed at Dover port on August 19, 1349, he proceeded directly to one of the royal residences to meet with Edward III. It had been the practice since the Norman Conquest of 1066 and establishment of strict royal control over the church in England—and confirmed by the papally approved Concordat of London of 1107 after a brief dispute over this important matter—that a bishop or abbot could enter his office only after he had sworn feudal loyalty (homage and feudal vassalage) to the king and received the "temporalities" or landed property tied to the episcopate or abbey directly from the king's hand, like any other major lord proceeding into his inheritance with royal approval.

Bradwardine went through this vitally important ceremony and received the enormous temporalities of Canterbury, the wealthiest bishopric in the kingdom, producing an equivalent of a billion dollars a year in revenue. Then he hastened to Rochester on August 21, to the palace of the bishop of that small diocese, the perennial official adjutant to the archbishop.

On the morning after his arrival Bradwardine developed

a high fever, which was thought to be the result of the exertion of his journey, remarkable by its length and speed for a man of fifty-nine, elderly by medieval standards. But on the same evening the dreaded buboes or black welts around the groin and armpits appeared, and his physicians acknowledged that the new archbishop was dying of the plague. He lingered for another five days and died on August 26. It is quite possible that a plague-carrying flea from a rat had bitten him on the ship while he was crossing the English Channel and that he was doomed before he set foot in England.

By this time the populace was weary of bodies struck down by the plague. It was feared that a mortally diseased body would somehow transmit the plague to anyone who approached it. In spite of this fear, Bradwardine's body was not immediately interred at Rochester. It was carried the twenty miles from Rochester to Canterbury and buried in the cathedral. His tomb is still there. That a risk was taken to transport a plague-ridden body a considerable distance signifies the affection and esteem in which Bradwardine was held.

The brand-new archbishop's death was a severe personal blow to Edward and the royal family, of whose entourage Bradwardine had been a part, especially after the recent death of young princess Joan. The plague was not sparing the royal family nor England's leading intellectual and religious figures.

Bradwardine's demise was that of a great public personage who exercised influence and did important things in many aspects of political life. On his way to Coblenz with the king in 1338 he had encouraged Edward to make a significant contribution to the building of Cologne cathedral. This monstrous pile survived the frequent Allied bombing of Cologne during World War II unscathed. It is the largest Gothic cathedral in Europe (the largest in the world, the Episcopal Cathedral of St. John the Divine is, strangely in Manhattan, New York, at the corner of Amsterdam Avenue and 113th Street).

In 1344 Bradwardine was representing his then-patron the bishop of Durham in parliament. He accompanied Edward through the great battle of Crecy in 1347 and the subsequent siege of Calais. Bradwardine's dispatches home carried some of the earliest news of these victories. Popular opinion attributed Edward's successes in part to Bradwardine's divine intercession.

Bradwardine himself encouraged this opinion. After Crecy he preached before the king and the royal family and entourage on the text from Second Corinthians: "God who always leads us in triumph grants victory to those whom He wills, and He wills to grant victory to the virtuous." In view of his visibility and impeccable reputation, Bradwardine's death in the plague was a psychological blow not only to the king and his family, but to political society in general. No one was safe.

Thomas Bradwardine remarked in one of his books that his parents lived in Chichester in Sussex in southern England. He probably came of old gentry or bourgeois stock. He entered Oxford University, then in its intellectually most flourishing era before the nineteenth century. To become one of the three thousand students at Oxford (in the mid-1950s, even after turning coed, Oxford had only seventy-five hundred students), he would have had at least nominally to become a clergyman, but this was a role that Bradwardine pursued professionally and sincerely.

He was a brilliant student among a group of the best philosophical and theological minds in Europe. By 1321, at age thirty, he was a fellow (don, faculty member) of a new residential undergraduate college endowed by a Scots lord especially for his young compatriots, Balliol College. By 1323 Bradwardine had an advanced degree in philosophy and theology and had joined the faculty of Merton College, the second-eldest (or perhaps first) of the Oxford colleges, as Master of Arts.

Merton College was the hub of a group of dons devoted to natural science, mainly physics, astronomy, and mathematics, in competition with another such group of intellectuals at the University of Paris. Bradwardine fell in with this mentor group and published a treatise on velocity that made him the intellectual star of the faculty. This was followed by a major treatise on theology.

Bradwardine's theorizing reached its high point in his arguing that space was an infinite void in which God could have created other worlds which He ruled as He did this planet. His treatise on astrophysics was not published in print until 1618, and its circulation in manuscript in the Middle Ages was a very small one. Therefore, his innovative doctrine on void and space was little known outside Oxford. But the impression Bradwardine made on his Oxford colleagues led to his recognition as a theorist of the first rank.

His theory that space was an infinite void in which God could have created other worlds had revolutionary implications. It departed from the comforting medieval assumption that this earth was God's only created planet. It looked forward to modern recognition that there may be other worlds on which life exists. Thereby it threatened the singularity of our world and of human life and its relationship to God.

Bradwardine did not develop the implications of his theory, probably because they would have struck at the center of medieval religion and moral belief. He was content to leave his theory in its astrophysical and less controversial form.

He had great ambition beyond academia. Becoming the center of a highly visible theological and moral controversy would have restricted his career possibilities.

Bradwardine was not content with his high academic reputation. By temperament, he was a social conformist. He wanted a grand career in the church and public life and advanced rapidly up the ladder of patronage, holding several ecclesiastical appointments until he became dean of St. Paul's Cathedral in London and confessor to the king and then for a couple of months the ill-fated archbishop of Canterbury. He artfully reached the top of the ecclesiastical pole, from which he was abruptly removed by the Black Death.

Heavyweight intellectuals rarely became archbishop of Canterbury in the Middle Ages. That does not mean the Canterbury archbishops were not well educated and highly literate in ecclesiastical Latin, a ponderous and difficult language (with the exception of Thomas Becket, a college dropout who couldn't read Latin and instead hired the best Latinist in England to translate into French).

The last front-rank philosopher and theologian who had been archbishop of Canterbury was St. Anselm in the early years of the twelfth century. He did poorly as archbishop, getting into needless quarrels with kings, exasperating the pope, and turning the monks of Canterbury into an ingroup of young gays.

What Bradwardine would have done at Canterbury is anyone's guess. But as a theorist he was formidable, and his theology related to the human response to the Black Death. God in His absolute and infinite being, thought Bradwardine, is totally beyond understanding. His actions whether of love or death cannot be rationally articulated by the human mind. God's predestination of human events can only be endured and blessed, not explained. But humanity has the rational capacity to begin analyzing and comprehending the natural world.

Eventually, seven centuries later, this philosophy would

generate modern medicine and its capacity to combat infectious disease, something beyond imagining in the time of the Black Death.

Bradwardine was the end product of the great Oxford intellectual movement that had begun in the 1240s and reached its end point with him and with an English Franciscan philosopher who had died of the Black Death one year before, in faraway Munich, Bavaria. This was another Oxford prodigy, William of Occam.

Occam was by 1348 an old man, in his mid-sixties, and he had not been in England since the second decade of the fourteenth century. Occam's disrespect for the absolute authority of the pope—or even of a general council of the church—inevitably attracted the attention of the papal court at Avignon and he was summoned there to be present during an investigation for heresy. As often happens today in Roman law countries, such as France or Italy, the investigation then conducted by a commission of cardinals and church lawyers was interminable.

After a couple of years Occam felt increasingly uncomfortable in Avignon, at once fanatical and corrupt, and he proceeded eastward and found refuge at the court of the erratic German emperor, Louis IV of Bavaria, who was at odds with the papacy over indefinable matters of church-state relations and prepared to offer protection to radical clerics like Occam.

In Munich at the emperor's court Occam was joined by

another provocative thinker and writer, an Italian-born professor from the University of Paris, Marsilio of Padua. As a graduate student, Marsilio had actually been elected rector of the University of Paris for a year in an ill-fated democratic experiment. Marsilio's sensational book, *The Defender of the Peace* (1324), called for subjection of the church everywhere, including the papacy, to the sovereign state. Only by this secular centralization could the peace of men be assured. So from somewhat different angles William of Occam and Marsilio of Padua were attacking the institutional power of the papacy in the church and society.

Bradwardine was not attuned to the Occamist-Marsilian frontal assault on the papacy. Whatever his opinion of what was going on in Avignon, he kept it to himself. In politics and society Bradwardine was a conformist and a conservative, deeply loyal to Edward III and his militarist state and at least outwardly obedient to the pope as long as the Plantagenet monarchy and the Avignon papacy maintained their somewhat tense but good formal relations.

Bradwardine was an intellectual partner of Occam in the realms of science, philosophy, and theology, but not in the theory of church organization. This intellectual movement was inaugurated at Oxford in the 1230s by Robert Grosseteste, who rose from being an orphan begging in the streets of Lincoln, to lecturer to the new group of Franciscan friars studying at Oxford, to the rich and powerful bishopric of Lincoln. In reaching this exalted level, Grosseteste's delicate service as tutor to the child king Henry III stood him in good stead.

Grosseteste was not formally a member of the Franciscan order of friars, but he was their official protector in England and sometimes wore their gray habit. In those days, Oxford did not have its own bishopric; it was part of the huge midland diocese of Lincoln. As bishop of Lincoln, Grosseteste was the chancellor of the university (a position equivalent to chairman of the board of trustees of an American university today). But he participated in the intellectual life of the university and wrote prolifically on a wide variety of philosophical and theological subjects.

He was also a pioneering experimental scientist (some historians say the first experimental scientist). His work on optics led to invention of corrective eyeglasses, a terrific boon to the near-sighted monastic scribes and university scholars who ruined their eyesight by working long hours by smoky candlelight.

The Order of the Friars Minor (Little Brothers) founded by St. Francis of Assisi around 1220, with a good deal of prodding and supervision from Rome, was supposed to be devoted to social welfare services for the poor and steer clear of the baggage of academic life. That was to be the mission of the other new Order of Preachers, the Dominicans. By 1260 the Dominicans had indeed come to dominate the graduate faculty of theology and philosophy at the University of Paris, through the prolific teacher and writer Albert the Great and his more famous and influential disciple, the Neapolitan aristocrat turned Dominican friar, Thomas Aquinas.

But under Grosseteste's leadership the English Franciscans became the vanguard for intellectuals at Oxford. By 1270 the two great northern universities were in competition—under Franciscan and Dominican aegis—for dominating progressive thinking within Latin Christian culture. Thomas Aquinas stood against the Oxford Franciscans who succeeded Grosseteste—Roger Bacon, Duns Scotus, and above all William of Occam—who took an opposite view of the relationship between faith and reason.

It was an intellectual split at the center of late medieval thought. The two sides continued their battles down into the sixteenth century. With Oxford's decline at the end of the fourteenth century—partly because it had become a hotbed of political as well as ecclesiastical radicalism and thereby antagonized the monarchy and nobility—the Occamist tradition migrated to the newly founded German universities. "I am an Occamist," said Martin Luther around 1510. He was an Augustinian friar teaching theology at a boondock new university in eastern Germany.

This was the intellectual world and academic milieu in which Thomas Bradwardine developed and rose to academic prominence. Unlike Grosseteste, Roger Bacon, Duns Scotus, and William of Occam, Bradwardine was not a Franciscan friar, a member of a religious order. He was an ordinary secular cleric, which gave him great intellectual freedom if at the same time denying him the protection and financial aid the Order of Friars Minor provided to its intellectuals.

Bradwardine had to make his way as a student and young don on his own, without special group support. This perhaps drove his precocity, although William of Occam was similarly an adolescent prodigy who started teaching at Oxford when he was barely twenty.

Bradwardine was intellectually as well as institutionally a free agent. He did not necessarily agree with the Oxford Franciscan tradition in all its doctrines and assumptions. But he was a product of the great Oxford intellectual renaissance of 1240–1380—not to be seen at the old English university again until the nineteenth century—and worked within the Oxford intellectual tradition as compared to the very different perspective of Parisian Thomism.

The Thomist school grew from the consequences of the penetration into the Paris University around the middle of the twelfth century of the cast corpus of Aristotelian science and philosophy through the medium of Arabic and Jewish schools (also writing in Arabic) in Spain and Sicily.

The Aristotelian corpus was translated in the eastern Mediterranean by Byzantine monks and Arab Muslim scholars into Arabic between 800 and 1000 and found its way, accompanied by various mathematical and medical texts, into Cordoba, Spain, and Palermo, Sicily, by 1050. Previously only Aristotelian logic was available in the West, which was thoroughly dominated by the Platonic idealistic and mystical rather than the Aristotelian scientific and rational frame of thought. By the middle of the thirteenth century Aristotle's writings were being translated directly from Greek into Latin rather than through Arabic mediation, and these improved translations were available to the corpulent and good-humored Dominican friar at Paris, Thomas Aquinas.

What drove the Thomist mission was a concern that Catholic doctrine was founded on the Bible, church authority, and the more mystical and irrational part of ancient culture, not on reason and Aristotelian science. It was to defend this established faith and high culture—something that the Cairo rabbi Maimonides had already attempted for Judaism, to deep resentment from the Orthodox rabbinate—that Thomas Aquinas, following his Parisian Dominican mentor Albert the Great, set out to show the large-scale compatibility of Catholic faith and Aristotelian reason and science.

Thus the Incarnation was based on revelation and historicity, but the existence of God could be proven by science and logic—there cannot be an infinity of causation; there has to be a First Cause in nature. A minimally decent Christian life—not one of saints and martyrs but of solid citizens—could be construed from Aristotelian ethics, with its advocacy of the Golden Mean and its insistence that "one swallow does not make a summer," that is, ethical behavior is habitually conditioned. The law of the state derived its legitimacy from the laws of nature (reason) and ultimately divine law.

In God's eyes and mind there was one truth about nature and man, Thomism claimed. Human reason could not fully establish this synchronous core truth but it could come close enough to qualm fears that faith and reason, science and revelation, were so separated and discrete that there was a "double truth," not a single one, as claimed by the radical Arabic philosopher Averroes and some of his followers at the University of Paris, such as Aquinas's adversary Siger of Brabant.

The Oxford Franciscan school, in whose wake Bradwardine followed, was having none of the Thomist synthesis. While not explicitly endorsing Averroes, they arrived at a similar position. They acknowledged that there existed a world of science, which could establish the rules of natural operation, preferably expressed mathematically. But it was superseded by a truth imparted by faith, an amalgam in varying proportions, depending on who was talking, of biblical revelation, church tradition and authority, and personal religious experience.

This is close to the modern position enunciated by Immanuel Kant around 1795 and which became the standard academic view after the intellectual and cultural wars over Darwinian evolution in the 1860s and 1870s. Kant's philosophy showed there was no other modus vivendi in university life and research.

Thus the Oxford approach of the early fourteenth century led ultimately to the modern scientific world, which, after about 1940 with the development of antibiotics, could actually counter an outbreak of infectious disease. Occam and Bradwardine were theoretically on the right track (along with an early-fourteenth-century Parisian master, John Buridan). But nothing tangible came of their remarkable scientific work, in which some historians have seen the clear beginnings of modern science, that was useful in any way against the Black Death.

In the first place, they worked only at physics and in that field their way was blocked until the sixteenth century by inadequate knowledge of algebra (derived from the Arabic world and ultimately from India). The only chemistry that was done was wasted on the dead end of alchemy. The way to biomedical science was blocked by reluctance to engage in dissection of the human body (created in God's image) and by the persistence of the medical theory propounded by Galen around A.D. 200 that good health was a matter of maintaining a balance of "humours" rather than combating specific disease-carrying microbes.

Since scientists had no microscope until around 1600 and no powerful one until around 1870, they could not see the disease-carrying bacilli. Therefore, in spite of some good work by Bradwardine and the Oxford school in the realm of theoretical physics (where it was relatively easy to catch Aristotle in error), the powerful and learned Oxford intellects had nothing to put forward to explain the Black Death.

Physicians attributed the plague to physiological imbalance, and when that story paled in the face of a raging pandemic, other explanations were trotted out. A commission of Parisian scholars assembled by the king soberly announced that the problem was astrological, something about Saturn in the house of Jupiter.

Of course moralists pronounced the plague to be divine retribution for sin, and while the sermonizers worked overtime to disseminate this conventional explanation, it was not convincing when the good and bad perished in equal numbers in the Black Death. Serpents and snakes were thought to be carriers of the plague, or it was attributed to Jewish malevolence.

In Germany in the fourteenth century it was widely claimed that the plague was the result of Jewish conspiracy—the Jews had poisoned the wells. This could not be a good explanation for the plague in England or France, however, because the Jews had already been expelled from England and from France, except Alsace, between 1292 and 1306.

The Black Death helped to make apparent that Thomism was an intellectual dead end. It failed to perceive the necessity for quantification in determining natural processes. It had no inkling of the crucial importance of experimentation. It was burdened with a strictly observational and rhetorical approach to science and furthermore remained specifically committed to Aristotle's error-driven physics.

Thomism looked liberal on the outside, a progressive philosophy that imagined a rationally constructed world. But Bradwardine knew the world was not rational. It was governed by an incomprehensible and awful deity whose actions, such as the Black Death, made no sense to humans.

The way forward to modern science and the medicine that finally conquered the plague was to accept emotionally, on faith, a fearsome and unpredictable deity of absolute power, who ruled this and possibly other worlds and spaces. Scientists then progressed through experimentation and quantification to the understanding of immediately complex, natural processes in very small segments.

This was the route advocated by Bradwardine, Occam, and the Oxford school. It led over long time to modern biochemically grounded medicine. Thomism, ordained in the sixteenth century as the official philosophy of the Catholic Church, led to liberal dogmas, happy dispositions, and intellectual nullity.

Thomas Aquinas was as learned an intellectual, as powerful a thinker and writer as existed in the medieval world. One segment of his doctrine, on the philosophy of law, endures today in what is called natural law theory and is still embraced by many liberal-minded professors in the better American law schools.

The intellectual road Thomas pursued—inaugurated by the Jewish thinker Maimonides in the twelfth century—seemed attractive and compelling at the time, but it did not lead to modern science. It did all the wrong things if that goal—which in the end distinguished Western European civilization from other cultures—was to be gained. It sought close compatibility between biblical faith and secular learning. It aimed at synthesis ("summa") of all important knowledge, while Galileo, Newton, and above all Einstein knew that truth was in the details, that knowledge of nature was gained by the closest possible scrutiny of very small segments of natural processes.

Instead Thomas decided that Aristotle was right about most things, including natural science. But Aristotle was wrong about many scientific things, as became evident to thinkers in the late thirteenth and fourteenth centuries, especially though not exclusively at Oxford University.

Bradwardine and his Oxford colleagues did not quite make the breakthrough to modern science. The quest had to be restarted in the seventeenth century, when algebra and scientific equipment were much more developed and the cultural ambience and academic reward system more propitious. But the archbishop knew the way to go, as Thomas did not: Focus on details, use quantification, do not try to force syntheses between science and theology. If the Black Death had not struck down the new archbishop, would the outcome have been different? Would the history of modern science in England date from fourteenth-century Oxford rather than from late-seventeenth-century Cambridge? The biographies of Sir Isaac Newton and Albert Einstein prove that a single great mind in a position of power and academic leadership can create an intellectual revolution.

Life and death in the era of the Black Death. A man is paying a prostitute while others are burning clothes in the hope of stopping the spread of the infectious disease. *Bodleian Library, Oxford, U.K.*

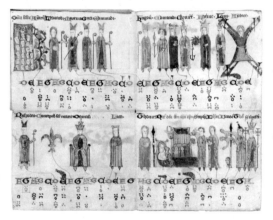

A procession of bishops depicted in the fourteenth-century saint's calendar. Episcopal prayers did not stop the Black Death. *Bodleian Library, Oxford, U.K.*

Bad news! The dead addressing the living, in a fifteenth-century manuscript. *Bodleian Library, Oxford, U.K.*

A mass for plague victims, as depicted in a fourteenth-century manuscript. *Bodleian Library, Oxford, U.K.*

A fourteenth-century depiction of male anatomy with the ever-present reminder, at the bottom, of the shortness of life. *Wellcome Library, London*

Carrying and burying victims of the Black Death. *Wellcome Library, London*

A mass grave for victims of the Black Death is located at Charterhouse Square, London. The buildings are postmedieval. In the fourteenth century this was the site of a Carthusian monastery, and victims of the Black Death were buried in a mass grave, which is now underneath the grass. *Anthony J. Gross*

The intercessionary religious procession of Pope Gregory the Great in 590, as depicted in a fourteenth-century manuscript, during the Black Death. This was the most immediate and common response to pandemic disease until recently. *Wellcome Library, London*

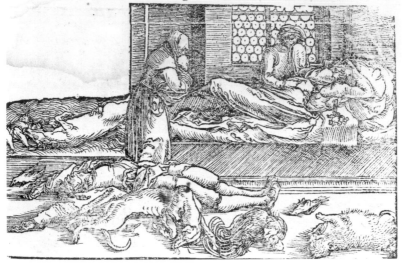

A sixteenth-century Petrarchan woodcut shows the Black Death as killing both humans and animals of many kinds as monks pray. The testimony to common impact of the Black Death on both animals and humans is biologically significant. *Wellcome Library, London*

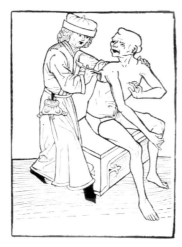

Lancing of the buboes, the black welts that appeared around the armpits and groin of victims of the plague, in a fifteenth-century woodcut. The remedy had little therapeutic effect. *Wellcome Library, London*

Doctors making theriac from snakeskin. Theriac was considered the all-purpose cure in the fouteenth century for the plague and other serious diseases. *Österreichische NationalBibliothek*

Astronomers looking at the heavens. Four-teenth-century knowledge of astronomy was considerable. It was used by French astronomers to determine the astrological cause of the Black Death. *Bibliothèque de l'Arsenal*

At Sedlec, in the Czech Republic, in the Cemetery Church of All Saints there is an ossuary (human bone depository) that contains thousands of relics of plague victims. In the late eighteenth century, a local woodcarver was given permission to create skeletal decorations from the ossuary deposits, resulting in a dramatic and little-known memorial to the Black Death. *Sandra Norman, Florida Atlantic University*

Passover plagues in a fifteenth-century Ashkanazi Haggadah. The ninth biblical plague is cattle disease and perhaps the origin of the Black Death. *British Library*

The burning of German Jews accused of poisoning wells and causing the Black Death. *British Library*

A 1492 depiction of the city of Strasbourg, where a massacre of Jews suspected to be responsible for the Black Death took place. *British Library*

Princess Joan left England with a musical introduction to Castilian culture—a Spanish minstrel—on her way to meet her fiance, Pedro of Castile. But the princess died in Bordeaux of the Black Death and never reached Spain or her prince. *Bodleian Library, Oxford, U.K.*

The Black Death forced an unprecedented exchange of fortunes. Shown above is Henry of Grosmont, duke of Lancaster. After his death of the plague in 1361, the duke's enormous estate passed to his surviving daughter, who was married to John of Gaunt, son of King Edward III. It became the foundation of John of Gaunt's wealth and power. *Crown Copyright, RCHME*

The suffering of Job in a fifteenth-century manuscript. As a result of the Black Death, Job was a figure that late medieval people could identify with. *British Library*

The Dance of Death in a late-medieval manuscript. Unsurprisingly, this was a popular motif in art and literature following the Black Death. *British Library*

CHAPTER SIX

Women and Men
of Property

❧❦❧

NINETY PERCENT OF THE wealth of England in 1340 lay in land. Of this land perhaps 40 percent was owned by the king and the royal family and the high aristocracy that usually carried the titles of duke, earl, baron, or simply "lord." Another 30 percent of the land was held by ecclesiastical officers and corporations. This left nearly 30 percent of the land to be owned by the rural upper middle class, who came to be called gentry in the fifteenth century. At most 2 percent was in the hands of free peasants, later called yeomen.

In England before the Black Death there were probably around half a million people in the gentry class, including women and children. By 1400 the gentry comprised half that number. Their family incomes varied as greatly as those of the American middle class today: anywhere from the equivalent of fifty thousand dollars a year in today's money to three or four million dollars a year. The lesser gentry were sometimes called esquires, an obsolete military term. Perhaps half of the upper gentry were called "knights," another obsolete military term. Knighthood entitled the senior male (as today) of the family to use the title of Sir and his wife to be given the honorific appellation of Lady. But a significant portion of wealthy gentry families resisted the official awarding of knighthood from the king because to be a "belted knight" could increase military and tax liabilities and put a heavier strain on the hospitality and entertainment budget of the family.

Marriage, the production of progeny, and inheritance were the core of gentry life. A gentry family in the fourteenth century could rise from relative mediocrity by favor of the king, by collecting booty in the French wars, or—less commonly—by careful "husbandry" or estate management.

But another and common route up the social and economic ladder was a series of good marriages—that is, those that brought in heavy dowries—plus the availability of male heirs steadily over several generations to keep the estate perpetually intact.

On the other hand, marrying a woman of modest means and thin dowry, or losing her rich dowry after her death because of complicated legal maneuvers by which most of the landed dowry was returned to her family, and absence of male heirs, or even widows who lived too long and sat on a big share of family income, could damage or ruin a great family. This relationship between generations and property was the central and certainly most interesting theme in the life of fourteenth-century gentry families.

Over this process of marriage, birth, death, and inheritance, the Black Death fell like a tornado sweeping across the countryside. It generated a much higher level of mortality than usual among the gentry—especially among male gentry—and many great families were suddenly shaken and their security threatened, their wealth and social status undermined. Coveted estates that had taken generations to build were suddenly swallowed by another family, distantly related, and the losing family's honorable name was expunged from society and history.

There were two kinds of people who especially benefited from the squabbles brought about by the Black Death and the endless litigation that was the result. The first were the "common lawyers" (so-called to distinguish them from civil—Roman—lawyers who practiced in ecclesiastical courts).

The common lawyers were graduates of the Inns of Court, the four residential law schools cum bar associations located in Westminster in London (they are still there today). They made their money protecting, expanding, and defending the gentry estates. There was actually a shortage of them, and their fees were high, but no gentry family could endure long without their services. Since medieval English procedure did not allow a defendant to be represented by an attorney in a criminal trial, the criminal justice bar was almost nonexistent.

Many of these barristers specializing in property and inheritance cases were on permanent retainer to leading families. It was not enough to know the law—not an easy thing to do because property law was frequently changed by judicial decisions as well as by occasional legislation (as in the United States today). They also had to be expert at drawing complicated documents in a highly specialized language, law French, and it also helped to know Latin and English, which were also used in the law courts. Above all they had to be expert "pleaders" (later called barristers), attorneys who were licensed to appear in court, stand on their feet, and with little or no aide-memoir argue immensely complicated cases before judges and juries for hours or days on end.

They perforce got so expert in doing these things that they created a body of real estate law that is largely still in effect today. A barrister of 1350 deep frozen and thawed out today would need only a six-month refresher course at a first-rate American law school to practice property or real estate law today. In every U.S. law school, a required course in the first semester for entering students is entitled "property." Its principles and procedures were worked out empirically by the English bar in the fourteenth century, given a big boost by the carnage and confusion visited upon gentry families by the Black Death.

The other beneficiaries of the plague, besides the lawyers, were women of the gentry class. The common law had a procedure for protecting widows, partly because the gentry landlords engaged in serial marriages with wives who died like flies in childbirth and were often gone by age thirty.

The heir to the family estate was usually a product of the first marriage and the widow, wife number two or three, was his stepmother, sometimes younger than the heir. Oedipal tensions ("that sexy young wench, my father's third wife and now his widow, is eating up the old man's estate," a great plot line that Shakespeare and Hollywood missed) could inflame an heir's greedy disdain for taking care of his stepmother or even his actual birth-mother, in this cruel, selfish society.

Therefore, the law stepped in and decreed that every widow had a right to "dower," one-third of the income (not the capital) of her husband's estate until she died. Within forty days of her husband's demise she was supposed to vacate the family mansion. But one-third the income from the family lands would allow her to live comfortably elsewhere and play the role of the grand dowager.

Complex legal instruments called jointures, a sort of prenuptial agreement, might allow her to recover part or even all of the landed dowry she had brought to the marriage, a common occurrence when the bride was much wealthier than the upwardly mobile groom. A great deal of late medieval litigation, miles of parchment court rolls, written on sheepskin, was taken up by litigation following the heir's visceral disdain for dower and jointure belonging to his stepmother or even his own birth-mother.

A favorite trick was simply to refuse the widow the entry

to the dower, leaving her in anxious frustration and genteel poverty. Then the heir's and widow's attorneys would try to work out a settlement giving the widow comfort but much less than she was strictly entitled to by law. If agreement could not be reached—and especially if the widow had important relatives who stood by her—then it was on to the courts where such cases could drag on not only for years but decades.

The worst thing that could happen to a gentry family was biological nullity—the family became extinct in children of either sex. No legitimate son or daughter survived the father or brother to inherit. Even without the added impact of the plague years, this occurred with remarkable frequency. In any given year before the Black Death, one out of twenty families of the wealthier gentry and also the nobility experienced extinction in direct succession.

If this happened a variety of outcomes could occur. The king might declare the family void and take the estate back into his demesne, his crown property, and then possibly give or sell it to an entirely unrelated family. Often a cousin would be allowed to inherit it from the extinct family, frequently taking their surname if he did not already bear it. This latter transaction required a hefty bribe to the royal government, in addition to the normal onerous inheritance tax, and could burden the estate for a generation or more.

Short of biological nullity, the worst thing that could happen to a gentry family was for two or three heirs in rapid succession, father, son, and even grandson, and all married, to die in a pandemic leaving three hale and hearty widows with dower rights in the family estate.

The fourth male heir (or rather his attorney, since he was likely to be a child) faced the gloomy prospect of two or even three dowagers asserting their dower rights upon the estate at the same time, taking away large parts of the family income and leaving a fraction only for the new heir. As the new heir came of age and found the majority of his ancestral income siphoned off by the two or three dowagers, personal animosity would exacerbate fiscal strain.

Then as now a dysfunctional family would be sucked into a whirlwind of psychological stress, fiscal tightness, and bitter litigation. It could take two decades to straighten out this mess, if it ever was straightened out. The lawyers clucked in pretended sympathy while adding up the steadily mounting legal fees.

No historian has yet come up with a statistical study of whether gentry males were harder hit by the Black Death than women in the same family. But there is plenty of anecdotal evidence to support such a theory. There was a plethora of dowagers in the two or three decades after the Black Death of the late 1340s. It is easy to understand how this difference in the mortality rate of male and female gentry happened.

The Black Death resulted either from bubonic plague or anthrax. The male gentry would commonly be out in the fields inspecting their lands, barns, and cattle and encountering plague-ridden rats or diseased cows daily. While there were some activist female gentry who would do the same thing, the majority of them lived a more confined and sheltered life and were less likely to have close daily physical contact with rodents and cattle.

The resulting difference in mortality from the Black Death was a boon to women of the gentry class. Their superior survival rate brought enhanced wealth, independence, and position in local society. But this sexual disparity could play havoc with the stability and economy of a gentry estate, especially due to the law's generosity to dowagers, and thereby generate decades of bitter family infighting.

❧

The case of the le Strange family demonstrates well what the Black Death and the courts could do to an unlucky upper-class family. The le Stranges lived in Whitchurch in Shropshire in the black-earth, high-grain-yielding country intensely competed for by gentry families. The rich le Stranges were ambitious and on the rise, and because of their upward mobility were starting to make marriages in some instances with younger daughters of the nobility.

But the le Strange family was exceptionally unlucky in losing male family members during three successive outbreaks of the plague—two in 1349, and one each in 1361 and 1375. By 1375 not even the relative fecundity of the family in producing sons for the next generation could help them escape extinction in the male line. The plague had eliminated sons and left ambitious dowagers.

The le Stranges going back to the 1330s were not originally a great gentry house. They were a family on the make, principally through marriages with rich women, plus good estate management. The enhancement of family fortunes was launched by the marriage of John le Strange the First to a wealthy gentry heiress, Anakretta le Botiler. In the next two generations the le Strange heirs married into the nobility. This raised their social and political profile and with luck would have accrued vast landed wealth to the family.

But the Black Death countered that luck. Fulk le Strange, John I's eldest son, married Elizabeth, the daughter of Earl Ralph of Stafford. Earl Ralph drove a hard marriage bargain. Fulk's father, Ralph Stafford insisted, had to settle land worth two hundred marks a year (about a half-million dollars) jointly on the couple. This meant that if both John I and Fulk died close in time to each other and Fulk's marriage to the heiress Elizabeth Stafford was short, the le Strange estate would be affected severely by loss of income from land held as dower for the widow.

Fulk le Strange died in the Black Death on August 30, 1349. But Elizabeth Stafford lived to a ripe old age by medieval standards, not dying until 1376. During those three decades Elizabeth not only collected dower from her deceased husband's estate but remarried twice, taking with her the succulent property that John I le Strange had to settle jointly on his son Fulk and Elizabeth Stafford to get Earl Ralph's permission for the marriage. The land thus eventually passed to the family of Reginald, Lord Cobham, Elizabeth Stafford's third husband.

The story gets worse and more complicated for the pathetic le Stranges. Not only did Fulk le Strange, the elder son and prime heir of John I, die in the Black Death in August 1349, but the old man himself, John I le Strange of Whitchurch, had died of the plague only five weeks earlier. For a rich gentry family this blow was equivalent to a 60 percent crash in the stock market today—if every single asset was held in stock.

Anakretta le Botiler survived her husband, John I le Strange, until the next visitation of the plague in 1361. This meant that there were now two living dowagers, Anakretta le Botiler le Strange and Elizabeth Stafford le Strange, both women from families powerful enough to get their full dower rights and then some. For the twelve years of her widowhood Anakretta held the family house at Whitchurch in Shropshire (contrary to custom, by which she should have vacated it within forty days of her husband's death). She held on to one estate that came with her dowry, since it was jointly visited upon her and John I. For another piece of land she paid her son John II le Strange and his estate the modest sum of twenty marks (thirty thousand dollars) a year.

This medieval soap opera in the age of the Black Death gets worse still for the le Strange gentry. John II le Strange got back some of his father's lands when his mother, Anakretta, died in 1361, but he himself died of the plague in the same year. This left a *third* dowager to be taken care of from the le Strange lands, a great lady indeed, Mary, daughter of the earl—later duke—of Arundel.

Mary Arundel le Strange had to be taken care of in the lifestyle she had come to expect as a product of the high aristocracy and as a lady dominating local society. She took possession of most of the income or actual real estate of the le Strange inheritance, dying in 1396. After the dowager Mary died, the remaining le Strange lands passed to Richard, Lord Talbot, who was married to Anakretta, the daughter of John II le Strange.

For Richard Talbot the benefits of this marriage alliance and ultimate inheritance were timely. His father, Gilbert, had run up huge debts and was sitting in debtor's jail in London. His debts were the result of a lifetime of unprofitable military campaigning in France and Spain.

Not all military gentry struck gold in looting the countryside and ransoming high-born prisoners during the Hundred Years War. Some put up their own money to campaign for the Black Prince or John of Gaunt and never got a return on their investment of time and money. Richard Talbot, newly enriched by the le Strange fortune, got his father out of debtor's prison and the old soldier died of the plague in 1387 in Spain, battling on to the end and losing.

The le Strange name thus disappeared from gentry history. Richard and Anakretta Talbot, however, turned out to be good managers in their efforts to recover the level of income, which had fallen during the great plague years of 1348–49, of their estate, Whitchurch. Some of this recovery stemmed from increasing commercial activity, which attracted tenants who hoped to get high market prices for their grains.

Overall, the experience of Whitchurch confirms the now well-accepted view of economic historians that in the agrarian sphere, the full economic impact of the plague was delayed by at least a generation. Even the fall in grain prices in the 1370s did not spell immediate disaster for careful landlords who were prepared to use imaginative means to boost their incomes. By the time of his death in 1396, Richard Talbot had actually increased income from the Whitchurch demesne by 25 percent and from the manor as a whole by 10 percent.

But successive outbreaks of plague and the failure of the population to make a sustained recovery made the situation gradually worse. Against this background, it was difficult for the manorial economy to bounce back from any new incidental blow that it suffered. There were several of these—a serious outbreak of plague in 1391, harvest failure in 1428, and floods in 1434.

But by far the most significant setback was the Welsh wars of the first decade of the fifteenth century. In the raids of 1404 Whitchurch was burned up to the gates of the manor by Welsh terrorists (or as they are called at Oxford today, freedom fighters). The damage was so severe that peasant rents were remitted for five years afterward.

The Talbots after about 1410 could not hire enough labor to continue direct cultivation of their own family lands, the demesne. They did what most other landlords did in the fifteenth century. They split up their demesne into farm blocks that they leased ("farmed") out to the wealthier and more enterprising peasants. From these yeomen leaseholders new gentry families emerged in the late fifteenth century.

Thus John I le Strange's great dream of a huge gentry estate and his family's eminence in local society over long time, and respect and honor for the le Strange name, came to absolutely nothing, even for the Talbots. The Black Death and all those privileged dowagers had brought this to pass.

In the high mortality of the Black Death there were plenty of other stories of women surviving and the men of the family perishing. This happened in Bordeaux in the family of the vintner and wine merchant Raymond le Clerque. In 1340 le Clerque drew up a long and complicated will dividing his land and property in and near the city among his six children, two sons and four daughters. His eldest son, Jean, got most of the property, and his second son, Guillaume Arnaud, one valuable vineyard. In return Jean was to pay five hundred pounds (about half a million dollars) to each of two of his four sisters, Margaret and Gaya, as dowries when they married. The other two sisters were apparently unprovided for.

Raymond le Clerque died in 1346. A year later the plague had completely undone his plans. In the end only Gaya survived, and the whole estate devolved on her and her new husband. Gaya became a grand lady. In 1351 she withdrew to a house in the more salubrious countryside, appointing her husband and a lawyer to manage her property in Bordeaux on her behalf.

Nor was the le Strange case singular in the way that the Black Death and bad luck in matters of inheritance could bring down a great family much as a massive stock market collapse could affect a very rich American family today. The story of the Hastings family, for example, reads like a comedy out of Shakespeare.

The tomb of Sir Hugh Hastings of Elsing in Norfolk (d. 1347) is representative in several ways of the world that was swept away by the Black Death. The style of the brass displays an aesthetic refinement that was not repeated in the aftermath of the plague. The brasses' iconography provides a history of family harmony that turned into bitter division as aristocratic estates passed between families more rapidly in an atmosphere of intensified competition.

The art historian Paul Binski has concluded that the brass on Sir Hugh Hastings's tomb was the product of an important London workshop, active during the 1340s and responsible for several impressive brasses. The style of this workshop is one of ebullience and confidence combined with artistic delicacy. It is distinguished by its eclecticism. Sir Hugh Hastings's tomb of 1347 is a delicate mix of English and continental influences, especially in the scene of the coronation of the Virgin, but also in the incorporation of an equestrian portrait of St. George above the figure of Sir Hugh.

This mix probably reflects the cross-Channel influences that transmitted to England through Edward III's French campaigns. Sir Hugh himself had fought at Crecy. The workshop flourished only briefly. After 1349 this style of brass completely disappears from England, leading scholars to believe that the artist had died in the Black Death or that the workshop had been broken up in the wake of the disaster.

Sir Hugh in his tomb is flanked by a group of eight weepers, headed by King Edward III and including companions in arms and members of the Order of the Garter such as Ralph, Earl of Stafford, and relatives such as Roger, Lord Grey of Ruthin. According to Binski, the tomb expresses "a form of corporate identity, embedded in shared military honor and badges of allegiance."

The Hastingses of Norfolk who commissioned this famous tomb were a younger branch of the family of John of Hastings, Earl of Pembroke, one of the richest men in England. The earl being childless, his nearest heir was his cousin Reginald de Grey of Ruthin, commonly called Lord Grey. Second in line was another cousin, William of Beauchamp.

As the earl of Pembroke prepared to leave anew for the continental wars in 1372, he had a bitter falling out with

Lord Grey. While the earl had previously been fighting in France, Grey dispensed a rumor that the earl had died abroad and proceeded to invade some of the earl's lands. When the earl returned he made Grey apologize publicly in front of several great lords but retained his pique against Grey.

Pembroke regarded Grey as a nouveau riche arriviste, and Grey was indeed descended from a younger son of a minor baronial family. The other cousin—and next in line as the earl's heir—was William of Beauchamp, descended directly from one of England's ancient families.

Before the earl again departed for the French wars in 1372, he placed his lands in irrevocable trust. This provided that in event of his death abroad the trustees were to grant the whole estate to William of Beauchamp, provided that the king would make Beauchamp earl of Pembroke. In effect John Hastings was adopting William of Beauchamp as his son and heir, with the king's anticipated approval and cutting out entirely Lord Grey. This was a clever but dicey legal maneuver, just the kind of complicated instrument a common lawyer loved to draw up.

The whole plan fell apart when John Hastings, earl of Pembroke, heard while he was in captivity in Spain that his young second wife was pregnant and then that she had borne him a son, John II. Pembroke had no doubt that he was the genetic father of the child. The trust that the earl had set up giving everything to William of Beauchamp had not provided for this possibility. In effect John Hastings had disinherited his own yet-to-be-born son by a rash and sloppy legal maneuver. "Everything that was done has gone wrong," he lamented.

John Hastings promised a monstrous ransom to his Spanish captors (equivalent to thirty million dollars) and hurried back to England to try to straighten out this mess. Between Paris and Calais in 1375 he took ill and died of the plague.

While Lord Grey was exulting that the trust meant to exclude him would have to be canceled—itself a very difficult legal engagement—the possible new heir, John II, made things easier for Grey by dying accidentally in a tournament. After protracted litigation that went on for two decades, Grey and Beauchamp settled the inheritance between them. Grey assumed the inheritance but by collusive prearrangement sold off a substantial part of it to Beauchamp at bargain rates.

The fears of John Hastings I, earl of Pembroke, of what would happen to his great estate and family name if it fell into the hands of parvenus had come strikingly to pass, partly because of the plague and partly because of his rash reliance on an expensive but incompetent lawyer who drew up a flawed trust.

At this point Edward Hastings of Elsing, descended from Sir Hugh, the knight portrayed in the magnificent brass of 1347, belatedly put in a claim. He went so far as to challenge Lord Grey to the archaic process (not finally abolished until 1819) of trial by battle: "Thou lying false knight,

I am ready to prove with my body against thy body," Edward Hastings proclaimed. Grey wisely declined the duel.

Edward Hastings's legal claim was very weak. The courts turned him down and ordered him to pay the enormous court costs of 987 pounds (two hundred thousand dollars). Though he could afford to pay he refused on the principle that he was the true earl of Pembroke.

For this ridiculous pride, Edward Hastings was consigned to debtor's prison and languished there for twenty years, complaining of the inhumanity that left him there "bound in fetters of iron more like a thief or traitor than a gentleman of birth." Not to say earl of Pembroke. After his wife died, he softened at last and made his peace with the now billionaire aristocrat Lord Grey, marrying his son to one of Grey's daughters. Edward Hastings died shortly thereafter. The Grey family played an important role in early-sixteenth-century politics.

Edward Hastings was undoubtedly a litigious crank, but there are some aspects of his irrationality that point up the subtle difference between the pre– and post–Black Death gentry worlds.

The plague had shaken the gentry society like an earthquake, and the fissures ran deep and long. It would be wrong to view the pre–Black Death era as a golden age of chivalry and consistently elegant behavior. But after the plague a certain restraining sense of honor and civility among the gentry and nobility was attenuated.

A seemingly endless war and the bitter politics of the reign of Richard II in the last two decades of the century certainly contributed to violence and rapacity among England's landed elite. The takeover of the crown by Henry IV of Lancaster, John of Gaunt's son, from the Black Prince's son, the gay and erratic Richard II, in 1399, Richard's condemnation as a tyrant by Parliament, and the murder of the deposed monarch, probably by starvation, were the ultimate symbolic acts of this new dark age of bad behavior.

The havoc randomly visited upon rising gentry families by the plague certainly contributed to the advent of an era of rural capitalism, unceasing aggressive litigation, and the conviction that unrestrained greed is good. American law students in their first-year course on property law are today imbibing a judicial heritage crystallized in Black Death England and the culture of contention and merciless conflict it embedded in the common law.

The social and cultural equation works in the opposite direction. If the judicial heritage of Black Death England embedded contention and merciless conflict in the common law and legal profession, the formality of the law itself and its slow-moving judicial procedures imposed restraint on the behavior of the gentry.

It is not that some wealthy gentry did not resort, like the great nobility, to gangsterism and violence. They did occasionally. But this was always measured against the due process of the common law and widely regarded among the upper middle class as bad behavior.

Under a renewed strong monarchy in the late fifteenth and the early sixteenth centuries the peaceful resolution of conflict and the juristic socializing of ambitions mitigated the violence and privatism and brought the gentry back to a rule- and process-driven life. It was not a generous or charitable culture that was transmitted into the eighteenth and nineteenth centuries, but it was one that operated within the rule of law. Within the juristic culture even dowagers had their share of triumphs.

⌾◍◍◍⌾

At first glance it seems that the gentry world in the age of the Black Death treated women of this class harshly—as property. But closer consideration moderates this easy judgment. Not only were widows privileged by the law of the dower, but brides bringing substantial landed wealth to their marriages were protected from abuse and impoverishment by prenuptial agreements giving them a joint ownership in the real estate that was the main part of their dowries, defending such wives from abusive treatment and curt dismissal.

Furthermore, even saying that married women in gentry society were mere property in the time of the Pestilence does not get at the reality of their situation. As in rich families today, the words *mere* and *property* did not go well together. There was nothing dismissive or pejorative about property. It was the heart and soul of gentry life. The males of the gentry household knew full well that their well-being and status followed closely from the level of property they held, particularly in landed estates. To be equated with property was no insult in the gentry world.

Nor did this mean that marriages were necessarily without love and passion, even if their transactional business character ultimately prevailed. The young people of this affluent class were well-conditioned in the psychology and ritual of romantic love. They were immersed in this culture, in the romances that they read or that were recited to them after dinner in the great halls of the stone family houses. They were prone and ready to take sexual initiative at any time after they reached puberty. Gentry women as yet did not wear underwear. Men wore a doublet with a movable codpiece covering their sexual organ. Coupling was quick and easy and the steady increase in the number of private bedrooms in the gentry houses—an amenity in the twelfth century reserved only for the head of household and his lady—facilitated sexual unions with due regard for female modesty, which wasn't very much to start with.

The one great lack in the lives of gentry women was their exclusion from higher education and the learned professions. It didn't inevitably have to be that way, but this sharp exclusion crystallized as the universities and secular professions, particularly law, were constituted in the period 1200 to 1350. This sexual segregation in higher education and the professions was not breached in England until around 1900, and not in substantial degree until around 1965.

In recent years there has been a tendency to lay the blame for medieval exclusion of women from higher education and the secular professions at the feet of church tradition and hierarchy. The church fathers St. Ambrose and St. Augustine, around 400, were outspoken misogynists, greatly respecting nuns for their spiritual qualities, but insisting that they be excluded from all leadership and sacerdotal roles in the church, and thus from the education needed to gain these posts.

Some historians have seen the triumphs of Ambrose's and Augustine's misogyny as the final chapter in a bitter conflict over the role of women in the church that goes back to the first century A.D. The attitude of the church fathers, further institutionalized by an exclusively male priesthood, can be regarded as the imposition of a male chauvinist position from which twelfth- and thirteenth-century ecclesiastical culture could not depart. Yet the church and medieval society changed in many directions over the centuries and their exclusionary attitude to women could have changed also. From the twelfth century on, many separated ("heretical") medieval religious communities, including the fourteenth-century English Lollards, allowed women preaching and leadership roles.

There was also a generation or so between about 1120 and 1160 when the church did produce top, highly educated intellectuals such as Heloise in Paris and Hildegard of Bingen in the Rhine valley, as well as a strong visible female hand in patronage of the arts and letters.

By 1250 the prospect for this road not taken was wiped away. The main reason was the inconvenience, instability, and costs that education and entrance into the professions would have meant for upper-middle-class gentry families. The English gentry families of the fourteenth century experienced no democratic ideological pressure toward enhanced privileges for their daughters. On the contrary, the weight of church tradition was strongly in favor of excluding women from higher education and the professions, and the gentry fathers and brothers had no hesitancy in cultivating these male chauvinist traditions.

There were frustrated intelligent and ambitious women among the gentry class of the late Middle Ages, of that we can be sure, who did not like early marriage and motherhood or the nun's veil and cloistered chastity as the only viable alternatives in their lives. Yet the great majority of gentry women of propertied families who followed the conventional role (and were still following it in the era of Jane Austen or even George Bernard Shaw and Virginia Woolf) still led comfortable and dignified lives, as much as women of the suburban middle class in America today.

The women to be pitied, if any, among the gentry class in the age of the Black Death were not married or widowed ones. The pitiable gentry women in the fourteenth century were daughters who for one reason or another—too ugly, too pious, and most commonly, lacking adequate dowries because the family already had too many other daughters, despite the widespread practice of female infanticide—found no husband and were shunted off to nunneries by the age of twenty. Some convents were still as most had been in the twelfth century—rich and genteel. The food in these high-toned establishments, nearly always of the Benedictine Order, was good. Diversion was gained by choral praying, writing, painting, embroidering, and (in spite of the indignant complaints of bishops) breeding and raising birds and greyhounds. When you win big at the greyhound track today, give silent thanks to those medieval dog-loving nuns.

In England of 1340, however, as was amply demonstrated as long ago as 1924 by the great medievalist Eileen Power, there were many dozens of small nunneries that were underfunded (funded by a rich family in the past and then forgotten about), even impoverished (sometimes by mismanagement), and sadly deficient in good food, entertainment, and amenities. This bleak ambience was unrelieved, in spite of the jokes and anecdotes about straying and licentious nuns—the medieval equivalent of *Playboy* magazine—by the sexual activity that married women and many a rich dowager enjoyed. They had to make do with piety alone.

CHAPTER SEVEN

The Jewish Conspiracy

━❦━

O N OCTOBER 30, 1348, AT Chatel near Geneva,
then part of the County of Savoy, a certain Jew
named Agimet, after twice being "put to torture
a little," as was allowed in continental Roman-based law,
made the following confession in a formal judicial proceeding before a panel of judges (translated by I. R. Marcus):

"To begin with it is clear that at the Lent just passed Pul-
tus Clesis de Ranz [a Jewish merchant] had sent this very
Jew, Agimet, to Venice to buy silks and other things for him.
When this came to the notice of Rabbi Peyret, a Jew of
Chambery who was a teacher of their law, he sent for this
Agimet, for whom he had searched, and when he had come
before him he said: 'We have been informed that you are going to Venice to buy silk and other wares. Here I am giving
you a little package of half a span in size which contains
some prepared poison and venom in a thin, sewed leather-
bag. Distribute it among the wells, cisterns and springs
about Venice and the other places to which you go, in order

to poison the people who use the water of the aforesaid wells and will have been poisoned by you, namely, the wells in which the poison will have been placed.'

"Agimet took this package full of poison and carried it with him to Venice, and when he came there he threw and scattered a portion of it into the well or cistern of fresh water which was there near the German House, in order to poison the people who used the water of that cistern. And he says that this is the only cistern of sweet water in the city. He also says that the afore-mentioned Rabbi Peyret promised to give him whatever he wanted for his troubles in this business. Of his own accord Agimet confessed further that after this had been done he left at once in order that he should not be captured by the citizens or others, and that he went personally to Calabria and Apulia and threw the above mentioned poison into many wells. He confesses also that he put some of this same poison in the wells of the streets of Ballet.

"He confesses further that he put some of this poison into the public fountain of the city of Toulouse and in the wells that are near the [Mediterranean] sea. Asked if at the time that he scattered the venom and poisoned the wells, above mentioned, any people had died, he said that he did not know inasmuch as he had left everyone of the above mentioned places in a hurry. Asked if any of the Jews of those places were guilty in the above mentioned matter, he answered that he did not know."

This detailed legal document (Agimet swore on the Torah

as to the truth of his confession) is accompanied by many other references to Jews under torture confessing to have spread the plague by poisoning wells, as in the following account from the Chronicle of Strasbourg in Alsace, a mainly German city that belonged to the French kingdom. It was also the only French province where Jews could still reside legally, having been expelled from the rest of France in 1306.

⊙∞∞∞⊙

"In the matter of this plague the Jews throughout the world were reviled and accused in all lands of having caused it through the poison which they are said to have put into the water and the wells—that is what they were accused of—and for this reason the Jews were burnt all the way from the Mediterranean into Germany, but not in Avignon, for the pope protected them there.

"Nevertheless they tortured a number of Jews in Berne and Zonfingen [Switzerland] who then admitted that they had put poison into many wells, and they also found the poison in the wells. Thereupon they burnt the Jews in many towns and wrote of this affair to Strasbourg, Freiburg, and Basel in order that they too should burn their Jews. But the leaders in these three cities in whose hands the government lay did not believe that 'anything ought to be done to the Jews.'

"However in Basel the citizens marched to the city-hall and compelled the council to take an oath that they would burn the Jews, and that they would allow no Jews to enter the city for the next two hundred years. Thereupon the Jews were arrested in all these places and a conference was arranged to meet at Benfeld [Alsace, February 8, 1349]. The Bishop of Strasbourg [Berthold II], all the feudal lords of Alsace, and representatives of the three above mentioned cities came there. The deputies of the city of Strasbourg were asked what they were going to do with their Jews. They answered and said that they knew no evil of them. Then they asked the Strasbourgers why they had closed the wells and put away the buckets, and there was a great indignation and clamor against the deputies from Strasbourg. So finally the bishop and the lords and the Imperial Cities agreed to do away with the Jews. The result was that they were burnt in many cities, and wherever they were expelled they were caught by the peasants and stabbed to death or drowned."

The town council of Strasbourg that wanted to save the Jews was deposed on February 9–10, and the new council gave in to the mob, who then arrested the Jews on Friday, the 13th.

The belief that the Jews were responsible for the Black Death first took root in southern France and neighboring Spain. In the fourteenth century there were only 2.5 million Jews in all of Europe, but a third of these lived in Spain and on the other side of the Pyrenees in southern France. The Jewish communities in this region were of long standing, in some parts of Spain going back to Roman times. They were relatively affluent, extremely literate, and in a relationship of growing tension with their Christian neighbors for both religious and economic reasons.

The rabbinical and capitalist elite in the Jewish communities, about 5 percent of the Jewish families, had furthermore come to abandon the Aristotelian rationalism of Maimonides, and instead embrace an esoteric theosophy called Kabbalah, which originated in the Hellenistic eastern Mediterranean in the first century A.D.

The Kabbalah intensified its mystical and astrological contents over time, its masters generating an air of mystery about themselves. Ordinary Jews were excluded from study of the Kabbalah. Only rabbinical families intermarried with the mercantile and banking elite were given access to it. Christians might well suspect that among the hermeneutic secrets of the Kabbalah were arcane recipes for magic, and poisons, and spells, that the Kabbalah constituted a kind of Black Magic.

The mid–fourteenth century was the beginning of the age of the witchcraft delusion that consumed Western Europe for another four hundred years. It was easy to somehow associate the Kabbalah and witchcraft and at the margin there may occasionally have been an actual connection. There is no doubt that there was a doctrinal overlap between the Kabbalah and the dualist Christian Catharist heresy in southern France, which so frightened and horrified church leaders in the early thirteenth century until the Cathars were suppressed by papal-authorized force.

If during the thirteenth century the rabbis had not withdrawn into the theosophic, astrological, and mystical shell of the Kabbalah; if they had remained loyal to the high intellectual road taken by Maimonides in the twelfth century, that of liberal rationalism and the effort to integrate Judaism with contemporary science, the way in fact taken by secular humanistic Judaism today, would that have made a difference in the treatment of Jews during the years of the Black Death? Would there have been less tendency for Christians to make scapegoats out of the Jews, charge them with spreading the plague by poisoning wells, and unleash horrible pogroms on them?

It is of course impossible to answer this counterfactual question. Since Heinrich Graetz wrote the first modern kind of history of the Jews in the 1870s, the overwhelming proportion of Jewish medieval historians have depicted medieval Christian treatment of Jews in the most negative manner. That the Jews were victims is clear, that the leadership of their intellectual elite might have made things worse has been underinvestigated.

The disturbing facts of the treatment of Jews during the Black Death remain. On May 22, 1348, King Peter of Aragon suppressed violence in Barcelona when twenty Jews were slaughtered and Jewish houses pillaged there. The wealthiest Christian burgesses tried to protect the Jews and launched a successful counterattack to repel the rioters. Royal officials also banned the preaching of inflammatory sermons in the city. Between May 17 and 19, there had been anti-Jewish riots in six additional Spanish cities.

The Jews in these places escaped by enclosing themselves within the walls of their quarters. All told, the Jews of Spain escaped lightly by comparison with those of northern Europe. Nevertheless in 1354, when an assembly of Jews met in Barcelona to form a band for the protection of Aragonese Jewry, the documents they drafted still lamented that many affluent and learned Jewish communities that had previously been safe and secure were destroyed suddenly. "A scattered flock of sheep is Israel," writes a contemporary Jewish writer.

Elsewhere, the attempts of the authorities to suppress violence and, in some cases, actively to protect the Jews began to meet resistance. In May Queen Joan of Naples had reduced taxes on the Jews on her Provencal lands by half in view of their losses in the riots. But in June her officials were expelled from Provencal towns. Pope Clement VI's bull of July 6, 1348, protected the Jews in Avignon and the vicinity but could not hope to be effective across the whole of Western Europe. (This is the same Clement VI who made fun of Thomas Bradwardine.) The pope was in fact a collector of Hebrew manuscripts. Nevertheless, the Jews became extremely vulnerable.

Secular rulers in various regions began to allow themselves to be swept along by the tide of anti-Jewish feelings. This began to occur in the Dauphine and in Savoy in southern France. It is likely that the limited power of the authorities in these regions made it unwise for them to attempt to contain the situation. The mountainous nature of Savoy made it an especially difficult region to govern other than by compliance with the will of the ruled, as is witnessed by its long history as a refuge for heretics.

Thus, in Dauphine when Count Hubert found that he could not keep order, he ordered the arrest of the Jews, while at the same time keeping up the pretense of pursuing those responsible for the persecutions. Between July and August pogroms spread throughout the country and Jews were thrown into the wells they had purportedly poisoned.

This was a critical stage in the persecution and not only because of the acquiescence of lay rulers. The plague moved more slowly through the foothills of the Alps, and news of its approach tended to run well before it. Two things resulted from this. The persecutions began to precede the advent of the sickness and there was time for legal processes—kangaroo courts that gave official and moral sanction to the atrocities, helping to counter the bull of Pope Clement, which had forbidden the killing of Jews without judicial sentence. Zurich in Switzerland banned its Jews in September 1348, about a year before the plague actually reached the city.

At Chillon near Geneva the attack on the Jews occurred four months before the plague arrived. This cause célèbre at Chillon was a turning point in the persecution. It is unlikely that the atrocities would otherwise have been so prolonged or intense. It was now that critical refinements were added to the story that the Jews had poisoned the water supply. The poison placed in the wells was supposedly made from the skin of a basilisk (a kind of mythical serpent), or from spiders, lizards, and frogs, or from the hearts of Christians and fragments of the Host (the wafer at the Mass).

At Yom Kippur, September 15, 1348, two Jews, Valavigny, a surgeon from Thonon (near Evian les Bains where the mineral water comes from today), and Mamson from Villeneuve (near Montreux, now a ski resort), were tortured into confession. Three weeks later Bellieta and her son Aquet were tortured, a process that was repeated ten days later until finally the tribunal had a thorough list of accusations against the families and their coreligionists.

This followed the confession of Agimet that he had allegedly gone on a poisoning tour at the behest of a Kabbalist rabbi. By November the blood libel and persecution had reached the upper Rhine, spreading into Germany. Stuttgart and Augsburg were among the first towns to be affected.

The escape of the Jews of nearby Regensburg, where 237 leading citizens formed a band to guarantee their protection, suggests that where the urban patriciate enjoyed some breadth of support among the artisans, it was still possible to quell mob violence. Compare this with Augsburg, where the burgomeister, Heinrich Portner, was heavily in debt to Jewish bankers and reputedly opened the gates to the "Jew killers," thus setting a pattern by which the rich and powerful acquiesced in the murders as a means of eliminating debts.

By December the murders began to take on increasingly macabre and outlandish characteristics. More and more Jews took their own lives rather than wait for their killers. At Esslingen in December they shut themselves in their synagogue and committed mass suicide by firing the building.

At Speyer the populace reveled in the novelty of encasing the bodies of their victims in wine casks and rolling them down the Rhine. "The people of Speyer," wrote a chronicler, "fearing that the air would be infected by the bodies lying in the streets or even if they burned them, shut them into empty wine casks and launched them onto the Rhine." We may ask how they could be sure that the cadavers had no poison secreted about their persons with which to infect the waters.

In Switzerland at Basel, the river was the scene of another tragedy. The town council made a noble attempt to ban some people notable for their ferocity against the Jews from the city. This increased the anger of the populace, which forced the repeal of the ban and the exile of the Jews, who were refused any possibility of return for two hundred years. To satisfy the popular mood some fugitives were rounded up and imprisoned until a house could be built for them on an island in the Rhine. There they were burned on January 9, 1349.

In Strasbourg on Saint Valentine's Day, a Saturday, the Jews were burned. According to a contemporary chronicler,

"They were led to their own cemetery into a house prepared for their burning and on the way they were stripped almost naked by the crowd which ripped off their clothes and found much money that had been concealed" (comparison with the Nazi Holocaust is obvious). The most reliable assessment is that of 1,884 Jews in Strasbourg, 900 were burned, the rest being banned from the city.

The populace then looted the synagogue, where they found the shofar, the liturgical ram's horn. Doubts about its purpose were resolved by the conclusion that it was a means of providing a secret signal to the enemies of Strasbourg to descend upon the undefended city and destroy it.

The conflagration continued during February in at least fifteen German and Swiss towns. When the Flagellants appeared on the scene they incited new persecutions. The Flagellants were a group of monks and fanatical laymen who believed that the plague was directly the result of human sin. They proceeded from town to town whipping each other and bystanders in the streets and causing general mayhem. Bishops hated them, but found it difficult to suppress them because the people took comfort in their displays of humility.

In July 1349, in Frankfurt, the Jews fired their own houses when they were attacked and set a large area of the city ablaze. The council of Cologne had resisted the persecution of the Jews despite the visit of a deputation from Berne bringing a Jew bound in chains who had confessed to his crime of poisoning wells.

The populace responded by attacking Jews in surrounding towns, including Bonn, where the civic authorities were less forceful. But the death of Archbishop Walram on August 14, 1349, made it more difficult to keep order and forced the council to give way to popular pressure. Some of the Jews killed themselves, while others fell at the hands of the mob.

At Mainz in the same month a riot started during a ceremonial of penance, when a thief stole a purse from the person next to him. The mob turned on the Jews, who took up arms to resist. Some of the rioters were killed (two hundred, in one estimate), and this inflamed the anger of the whole community. When the Jews were finally overpowered, they set fire to their own houses, creating such heat that it melted the bells and the lead in the windows of the Church of Saint Quirn.

East of the Rhine the attack on the Jews gradually petered out. Two of the principal rulers, Albert II the Wise of Austria and Casimir II of Poland, were determined protectors of the Jews. However, not even these princes could completely stem the tide of persecution.

Jews often traded in spice or as apothecaries and many practiced as doctors, which brought down additional suspicion on their heads. But the same expertise could bring them protection from the educated and powerful, who valued their services, though this seems to have been far more effective in southern Europe than in Germany and the north. The personal physicians of both Pope Clement VI and Queen Joan of Naples were Jewish.

Part of Pope Clement VI's favorable disposition to the Jews undoubtedly stemmed from his respect for Hebrew learning and for scientific expertise in particular. Jewish Kabbalistic mastery of astrology also made a favorable impression on the Avignon pope.

But despite his personal sympathies and bulls forbidding persecution, there were more general ways in which Clement helped to sustain the concept of Christian society that promoted persecution and expulsion. He believed as strongly as any pope in the principle of preserving the church from corrupt influences through expulsion and isolation.

In a vituperative sermon preached against Emperor Louis of Bavaria in 1346 he had declared the ruler "a putrid and infected member . . . a rabid dog, a cunning wolf, a fetid he-goat and a cunning serpent." Expulsion from Christian society was the only appropriate recourse.

For Pope Clement, who copied the Hebrew alphabet into his commonplace book, Jewish and Christian learning were part of the same organic whole. Christendom had issued from Israel and the Jews should not be anathematized and extirpated in the same way as heretics. Such a subtle, academic distinction was bound to carry limited force.

The suspicion that powerful individuals and Christian communities as a whole cynically used the plague as an opportunity to dissolve their debts to the Jews and to recoup some of the wealth that had passed into Jewish hands has long hung over the plague pogroms. The story is in fact a complex one. The Jews were relatively small-time lenders, which made them useful to great princes, who preferred to deal with them rather than the more powerful Florentines and Lombard bankers. But the Jews were detested by a wide spectrum of local society, from minor princes to artisans and peasants.

It is true, however, that there were large amounts of money to be made from the destruction of the Jews. In Cologne the razed Jewish quarter became the linchpin of the property empire of Arnold of Plaise, one of the greatest land speculators in late medieval Germany. The archbishop and town council, who shared Jewish property within the walls, spent part of the windfall beautifying the cathedral and building a new town hall in the Flemish style. That does not necessarily mean we should interpret the disaster as simply a premeditated act of profiteering.

Some of the stories of individuals' debts and debtors reflect the complexity of the situation. The most notorious case is that of the Augsburg burgomeister Heinrich Portner, who reputedly opened the city gates to rioters pursuing the Jews. Contemporary deeds from Regensburg show that Portner was genuinely in debt to Jewish lenders. In 1345 he had borrowed money at 25 percent. But this does not entirely provide a credible motive for the actions attributed to him. The interest rate, though high, was not excessive by the standards of the mid–fourteenth century.

More emotive than pure cash were the regalia (symbols of

office and religious artifacts) that rulers and bishops some-times pawned in order to raise funds. The Jews' preparedness to handle these items easily touched the Christian unease over the relationship between their monetary and sacred value.

The bishops, in particular, have been thought cynical and greedy in their treatment of the Jews. While this view is justi-fied to some extent, the position is complex, and attitudes varied dramatically between individual bishops. Archbishop Baldwin of Treves protected Jews under his direct authority and even demanded that the Strasbourg council return the properties of Jews that had been seized. But he was neverthe-less in a position to profit from the persecutions.

On February 17, 1349, Philip VI, the French king, gave Archbishop Baldwin all the properties of the Jews who had been slain in Alsace and elsewhere or might yet be slain. Ma-jor sums had to be expended on securing election to episco-pal sees and the Jews represented an obvious source from which these debts could be recouped. Archbishop Berthold of Strasbourg began his persecution of the Jews in precisely this way. But few clerics were as single-minded in this matter as he was.

Archbishop Walram of Cologne owed five hundred gold pieces (at least a million dollars) to the Jewish banker Meyer von Sieberg. In 1334 Walram had his creditor arrested and executed for abetting counterfeiters. His terrified widow, Ju-dith, renounced the archbishop's debt. When the Black Death persecutions reached Cologne, Walram's authority helped to keep the mob pursuing the Jews out of the city, diverting it into the towns and villages of the surrounding countryside.

It was only Walram's death on August 14, 1349, that exposed the Jews of Cologne itself to annihilation. After the elimination of the Cologne Jews the archbishop was declared heir to all their property outside the city walls and half that was within. It did not necessarily suit the great ecclesiastics any more than other rulers to eliminate the Jews altogether. They were too useful as lenders who could also be heavily taxed in return for protection. Beyond this, the great magnates who served as bishops had a predisposed dislike of popular rioting (even anti-Jewish rioting), which could turn into a general peasant or artisan revolution.

It is possible that popular belief that Jews were responsible for the Black Death in France and Germany was inspired or at least exacerbated by a visibly lower incidence of the plague among Jews.

By the mid–fourteenth century a variety of strictures against Jewish engagement in farming had been imposed as the Justinian Code of Roman Law became after 1150 the basis of the continental legal systems. Incorporated in the Justinian Code were imperial edicts from the late fourth century that drove the Jews off the land by prohibiting them from using Christian labor.

By the middle of the fourteenth century the now almost exclusively urbanized Jews were being segregated into distinct quarters within cities, often called the Old Jewry, and after 1500 the ghetto (which originally was the insalubrious section of Venice near iron foundries where Jews were assigned). By the time of the Black Death, Jews had also lost the place as great international merchants that they had enjoyed around A.D. 1000, giving way to Italian merchants in particular.

Segregated in their own quarters (where the Jews lived as humble artisans except for a very small elite of bankers and rabbis), the Jews were cut off from the rodents on the wharves and the cattle in the countryside, the main carriers of infectious disease. In addition, rabbinical law prescribed personal cleanliness, good housekeeping, and highly selective diets. These conditions may very well have isolated the Jews from the hot spots of plague and their practical quarantine aroused suspicion that they were responsible for the disease to which they themselves seemed immune. Of course only a general paranoid attitude to Jews among the populace could activate these suspicions into pogroms.

The Black Death pogroms against the German Jews had the inevitable effect of making them feel frightened and insecure. When Duke Casimir II of Poland not only tried to protect Jews in his domains from pogroms, but invited Jews to move eastward and settle in his vast, underpopulated domains, large numbers of Jews began to move en masse to Poland.

This immigration continued into the sixteenth century.

Like many Western European rulers of the early Middle Ages (700–1000), the Polish duke and his successors saw the Jews as an economic asset, bringing credit facilities and long-distance trade to the country.

By 1500 the Jews had been assigned an additional role of importance in Polish society and the frontier Ukrainian lands also ruled by the Polish nobility. They were widely employed as estate agents for the Polish nobility, supervising thousands of peasants forced into serfdom and managing the exploitation of the rich Polish and Ukrainian soil. Jewish males became trilingual—Hebrew for liturgy and rabbinical learning, a Slavic language for business, and Yiddish, a late medieval German dialect written in Hebrew characters, for everyday life in their own communities (most Jewish women knew only Yiddish).

By the mid-sixteenth-century Jews were rewarded for their services as estate agents with a lucrative monopoly in selling liquor to the peasants. This is the origin of the Yiddish folk song "a Gentile is a drunkard." Jews also prospered as lumber and fur merchants. Great schools of rabbinical learning, many still in existence when night descended in September 1939, emerged in Poland and the Ukraine. By the early seventeenth century half of the Jewish world population of 3.5 million lived in Poland and the Ukraine.

The Jews came to love the Polish and Ukrainian physical environment and in the nineteenth century (if not much earlier) wrote poetry lavishly praising the farmland, forests, and climate of Eastern Europe. The rise of the great Jewish communities in Slavic Europe, remarkable for their enterprise and traditional learning, and also innovative in religious and literary expression, was a direct result of the Black Death.

The prosperity and security of the Polish and Ukrainian Jewish communities reached their height in the first half of the seventeenth century. In 1648 the first great reversal of Jewish fortune occurred. A Cossack and peasant rebellion against the Polish landlords directed itself against the nobility's Jewish estate agents.

In the second half of the eighteenth century a long-term deterioration of Jewish living standards began that in effect was never reversed. The problem of overpopulation in the Jewish villages and towns, now often controlled by reactionary Hasidic rabbinical families, was exacerbated by the incapacity of the Polish nobility to maintain their economic status and political independence.

In the 1790s, with the partition of Poland, 75 percent of the Eastern European Jewish population passed under the rule of the czarist Romanov empire. Infected by the intense anti-Semitism of the Greek Orthodox Church, the czarist government and the Jewish rabbinical leadership were not able to work out a modus vivendi that could forestall increasing Jewish poverty. The Romanov government divested Jews of their liquor monopoly and made it very difficult for them to migrate eastward into Russia.

In the thirty years before World War I there were some

improvements in Jewish economic conditions with the penetration of the Industrial Revolution into Jewish cities in Poland, the Ukraine, and Belarus. A secular Yiddish culture of extraordinary intellect and vitality emerged in Odessa and other centers. But this progress was arrested by the war and the postwar hostility of the Soviet commissars, many of them initially products of the Jewish revolutionary left, to Jewish religion and identity.

The new Catholic state in postwar Poland was hostile to Jewish prosperity and civil rights. Then came the Nazi invasion and the Holocaust of the Jews with extensive assistance of Poles and Ukrainians.

Since 1945, because of the centrality of what in 1948 became the State of Israel in Jewish life almost everywhere, Jewish history has been written through the prism of that little strip of sandy land on the eastern shores of the Mediterranean that has also variously been called Canaan, Judea, and Palestine. With the extermination of the 3.3 million strong Jewish community in Poland and decimation of the Jewish populations in Lithuania, the Ukraine, and Belarus during the Nazi occupation, the use of Zionism and Israel by the rapidly assimilating American Jews to latch on to a dissolving transgenerational identity has become ever more pronounced in recent years. The disappearance of a generation of Yiddish speakers has mightily contributed to this development, even though a handful of Jewish studies programs have made marginal efforts to resurrect Yiddish language and culture.

Yet it was in Poland and its frontier territory of the Ukraine in the period from about 1480 to 1640 that Jewish thought and culture as it flowed into the early twentieth century was truly molded. It was a world whose language, communal organization, and religious movements were a legacy of late medieval Germany from which the Black Death and its attendant pogroms had impelled the Jews eastward into Slavic domains.

Only in America and Canada at the end of the twentieth century would Jews again attain the level of security, prosperity, and learning that they enjoyed under Polish monarchy and nobility in 1600. Poland was for them the golden *medinah* (country). There they created a distinctive thought world and behavior pattern that was still central to Ashkenazi (European, Western) Jewry in 1900. The casings and reverberations and genetic base of this post–Black Death Jewish world of Eastern Europe endures without the Yiddish language base and with much of its history forgotten in Anglophone North America and Hebrew-speaking Israel today.

The huge migration of Jews from the former Soviet Union in the late 1980s and 1990s to Israel and the United States has subliminally and genetically reinvigorated this bond with the Slavic past of the sixteenth century even though under Stalinist repression the Soviet Jews lost much of their linguistic and cultural identity. Perhaps if we wait long enough history will surprise us.

PART III

HISTORY

∽

CHAPTER EIGHT

Serpents and Cosmic Dust

IN THE OUTBREAK OF A pandemic there is, as Richard Evans, the historian of disease in nineteenth-century Hamburg, has suggested, a common dramaturgy. This involves flight to supposedly safe areas, usually in the putative bucolic countryside, and a conspiracy theory blaming the disease on strangers and unpopular minorities. As late as 1831, an outbreak of cholera in Poland, caused, as always, by polluted water, was attributed to Jews.

Modern medical science has reached a consensus that the Black Death was mainly bubonic plague spread by parasites carried by rodents, chiefly rats. Another infectious disease, anthrax, a cattle epidemic (murrain) was also involved, spreading from very sick cattle to humans. A consensus can be wrong. Over the years explanations of the Black Death have varied greatly and passed in and out of style. The earliest medieval explanation for the Pestilence was reptiles; one of the most radical contemporary ones is cosmic dust.

Today the animals associated with the plague are rats and

fleas, but these did not figure strongly in the medieval imagination. Neither of them feature in English medieval bestiaries or animal encyclopedias, though mice and ants do. Although William Langland's *Piers Plowman* (c. 1380) portrays the rat as an undesirable creature, this is because it devoured men's food supplies, not on the grounds that it spread disease. The absence of references to rats as a necessary prelude to a human epidemic may arise from contemporaries' failures to notice or attach any significance to the rat plague. There were, however, contemporary suggestions that domestic animals, such as cows, could fall sick of the plague.

Contemporary society looked toward less familiar beasts dwelling in remote areas that could be identified as the sources or the companions of virulent poisons that had allegedly contaminated the entire atmosphere. Much emphasis was placed on the sea as the source of the infection. Louis Sanctus of Beringen, writing from Avignon on April 27, 1348, says that the infection arrived in Europe by three galleys that arrived at Genoa "carrying horrible disease from the East." Expelled from Genoa after locals fell sick, the ships went from port to port carrying disease wherever they went, until one of them reached Marseilles.

This story may of course be the literal truth, but the story has powerful resonances with literary motifs, such as the ship of fools and the ship of the damned. Beringen goes on to say that people refused to eat spices, which might have arrived on infected galleys, and also certain sea fishes, presumably because they were thought to come from the contaminated oceans.

A more important theory about the origins of the plague's poison, and one that must be considered in the context of the more widespread concept of poisoned air or miasma, concerned exotic reptiles in general and snakes or serpents in particular. Beringen recounts how in September 1347 in a certain province in the Orient, there had been three days of calamities that had terrified the whole country. On the first day there had been a rain of frogs, serpents, lizards, scorpions, and many other kinds of venomous animals. All of these are frequently illustrated in English bestiaries where, interestingly, the scorpion is represented as a kind of lizard. Other contemporary writers make very similar reports, speaking of rains of fire and of venomous creatures, including plagues of snakes and scorpions that had supposedly occurred in countries between Persia and China. It was also suggested that the reptiles had been released from below ground by earthquakes along with the corrupt air that caused the disease.

These stories recalled the biblical pharaonic plague, preceding the deaths of the firstborn. The magical significance of the snake is also reflected in Moses' transformation of Aaron's staff before pharaoh.

From a practical point of view, the most significant feature of the snake's supposed power was that it was curative as well as infective. Snake poison, blood, and roasted, dried snake meat are still important ingredients in Eastern medicine. In medieval Europe theriac, or treacle, was the most prized and the most expensive of all medicines, and although the recipes for it are numerous and various, the common content was always mashed snake's flesh, which had often been skinned and roasted before being left to mature for a year or more.

In England, where most theriac arrived in prepared pots carried on Mediterranean galleys, the Grocer's Company guild kept a careful eye on what was a lucrative luxury trade that was especially open to abuse. The disruption of trade by war and the increased demand in southern Europe may have caused a shortage of the drug in England in 1349, at the very time when it was most sought after.

The basis of theriac's efficacy was commonly held to be homeopathic—one of its primary uses was as a remedy for snake bite. In his book on medicine (a book about spiritual remedies for sin that sheds interesting sidelights on medical practice), Henry of Grosmont, duke of Lancaster, who would himself fall victim to the plague of 1361, wrote that "the treacle is made of poison so it can destroy other poisons."

Grosmont also thought of theriac as a moral curative. It was the medicine "to make a man reject the poisonous sin which has entered into his soul." Since the plague was held to be a disease imposed by God as punishment for sin and had its origins in poison associated with snakes and reptiles, theriac was a particularly appropriate remedy or prophylactic.

One fifteenth-century English book of medical advice suggested that during times of pestilence theriac should be taken twice a day dissolved in clear wine, clear ale, or rosewater, but well before meals so that it had time to take effect.

During the initial outbreak of the Black Death, the greatest advocate of theriac was the eminent physician of Bologna and Perugia Gentile of Foligno, who himself died in June 1348 of the plague. In his plague treatise Gentile stressed that the theriac used should be at least a year old. Children, he thought, should not ingest it but should have it rubbed on them instead.

Gentile also counseled other remedies for these reptilian poisons. One of his cordials incorporated emerald, a stone he believed would crack the eye of any toad that looked at it. In addition, he recommended the use of an amulet ring, the specifications of which had descended from the kings of Persia. An amethyst was to be inscribed with the figure of a man bowing girded with a serpent whose head he holds in his right hand with the tail in his left. The stone was to be set in a gold ring with the root of the serpent encased beneath it.

Despite the popularity of theriac, it was a controversial medicine because of its almost universal application, magical associations, and the difficulty of producing a rational analysis of its efficacy consistent with the current intellectual system. This was important because the snake was the symbol and alternative shape of Asclepius, the pagan god of healing whose cult, the most enduring of those of the ancient gods, had been bitterly opposed by St. Augustine in the early fifth century. This tradition is still reflected in the Hippocratic staff, the symbol of the medical profession.

Nicholas of Poland (fl. 1270) spent twenty years at the medical school of the University of Montpellier, where attempts to find rational causes for the value of theriac were among the chief preoccupations. In the end Nicholas rejected this theorizing and returned to Poland professing its irrelevance. But he was utterly convinced of the therapeutic value of snake, toad, and lizard meat. He thought that kings and noblemen, who could afford it, should eat it at every meal. God, according to Nicholas, had conferred marvelous virtues on all of nature. "The more filthy abominable and common things are," Nicholas argued, "the more they participate in these marvelous virtues."

This contrasted with the employment of medication based on gold, advocated by doctors who favored alchemy. Nicholas believed that these doctors were quacks who were robbing their patients of their money and hastening their deaths. Of course, most doctors at the time of the plague, like Gentile of Foligno himself, used gold as well as theriac in their treatments. But for Nicholas the difference was that he believed that theriac had been proven effective by experience.

The historian of early medieval medicine Peregrine Horden adds more to the snake theory. He says that in the medieval mind mythical snakes, serpents, and dragons are to a large extent interchangeable, and that dragons are characterized by poisonous, pestilential breath.

Bishop Gregory of Tours recounted that in November 589 the Tiber had flooded the city of Rome. "A great school of water snakes swam down the river to the sea and in their midst was a tremendous dragon as big as a tree trunk, but these monsters were drowned in the turbulent waves of the sea and their bodies were washed up on the shore." There then followed a plague epidemic of which the first victim was the pope, Pelagius. Gregory the Great, his successor, then led the Roman populace in penitential processions. What else could the new pope do?

Gregory of Tours noted torrential rains as a sign of the arrival of plague in Gaul. Gregory of Tours' account resonates with the recent book *Catastrophe*, by David Keys. In this version a tremendous volcanic eruption in Java in 535 had extremely disastrous results all over the globe, generating monstrous rainstorms and flooding that was somehow connected to the outbreak of plague in the sixth century.

Gregory of Tours may have been on to something, and the medieval connection of the plague with sea serpents could have been a memory of what had occurred in the sixth century. Today there continues to be association of floods with infectious disease on the epidemic scale. When we are shown devastating floods on the TV news, whether they occur in North Carolina or Mozambique, the TV reporter intones that not only have these poor people lost their homes and their lives, but there is now widespread fear in the flooded area of aggravated infectious disease.

What may be happening in these instances, operative in the world today as in the Middle Ages, is the impact of a Jungian archetype. Flooding equals serpents equals a pandemic. The TV anchors are functioning within the same universal archetype as was Bishop Gregory of Tours in the late sixth century.

∽

Diseases coming to earth from outer space can be viewed as another Jungian archetype, in modern times worked into redundancy by science fiction writers. But this banality does not rule out a real scientific basis for the idea of diseases from outer space.

The theory that the Black Death originated in outer space dates back to a book published in 1979, *Diseases From Space*, by Fred Hoyle and Chandra Wickramasinghe. Since then the two authors have published a series of books, one as recently as 1993 (*Our Place in the Cosmos*), in which they have responded to new developments in research and, to some extent, to criticism of their thesis. Hoyle is a renowned Cambridge astrophysicist and Wickramasinghe is Professor of Applied Mathematics and Astronomy in the School of Mathematics in the University of Wales, Cardiff.

Hoyle and Wickramasinghe argue that contrary to what is usually believed, conditions on primeval earth were not suitable for the production of life. It is far more likely that life arrived on earth from somewhere beyond it.

Darwin's theory of evolution works up to a certain point, say the two scientists, but there are worrying inconsistencies. Plants and animals show evolutionary characteristics that bear no obvious relation to their chances of survival. Some bacteria exhibit resistance to high levels of radiation and cold far more extreme than that experienced anywhere on earth—the qualities that would equip them for survival in space.

Interstellar space is full of seeds or grains that have the characteristics of hollow dehydrated bacteria. The conditions of interstellar space could not permit the replication of organic material but this is not true of comets, which contain large quantities of water and, due to evaporation, expel large quantities of dust when they pass close to stars.

Experiments conducted when Halley's Comet passed close to the earth in 1986 showed that the materials it expelled closely resembled the "organic" grain found in interstellar space. The result of this emission of matter from comets is that "an immense number of bacteria and viruses of all kinds fall each year from space onto the terrestrial ensemble of plants and animals."

For Hoyle and Wickramasinghe the history of human disease is a kind of "pathogenic test" of this rain of organic matter from space. In their view all diseases ultimately have an extraterrestrial origin, what they call "vertical transmission." They do not rule out, however, the further passing of an illness from person to person, "horizontal transmission."

Once introduced, a vertically transmitted disease can establish reservoirs of contagion, which can sustain it for a period, potentially even for centuries, without further intervention by nonterrestrial sources. Tuberculosis is seen as such an ever-present disease. Other diseases establish reservoirs but diminish in potency until boosted by a further fall of extraterrestrial pathogens. Smallpox is an example of such an illness.

According to this theory, bubonic plague is a prime example of a disease chiefly attributed to vertical transmission, because it "has appeared in sudden bursts separated by many centuries and it is difficult to understand where the plague bacillus, *Y. Pestis,* went into hiding during the long intermissions." They concluded that after weakening over time, the bacterial reservoir disappeared. But the bacterium *pasturelle pestis,* bubonic plague, persists as an irregular visitor to our planet.

According to Hoyle and his associate, after A.D. 600, "Bubonic plague seems to have disappeared from our planet for eight centuries, until it reappeared with shattering personal and social consequences in the Black Death of 1348–50." After smoldering with minor outbreaks until the mid–seventeenth century, it disappeared again for two centuries before recurring in China in 1894 and spreading to India.

Hoyle and Wickramasinghe support their case by shrewdly pointing up some inconsistencies in the conventional explanation of the spread of the plague. Although they accept that plague involves a rat disease, they are scathing about the view that the disease could have been spread across Europe exclusively by migratory rats.

"To argue that stricken rats set out on a safari that took them in six months not merely from southern to northern France but even across the Alpine massif, borders on the ridiculous. What remarkable rats they were! To have crossed the sea and to have marched into remote English villages, and yet to have effectively bypassed the cities of Milan, Liege and Nuremberg," where the incidence of plague was very low. (Milan, it may be noted, enforced a quarantine that may have saved its citizens from the plague.)

The absence of plague in Bohemia (today the Czech Republic) and Poland is commonly explained by the rats' avoidance of these areas due to the unavailability of food the rodents found palatable. Instead of the alleged finicky diet of Central European rats, Hoyle and his associate point to a climate that was not propitious for the sprouting of disease pathogens from outer space.

In conclusion they argue that this generalized spread of the Black Death with exceptional gaps is entirely consistent with a fall of pathogens from space. "There was no marching army of plague stricken rats. The rats died in the places where they were."

In spite of Hoyle's and Wickramasinghe's impeccable scientific credentials, their thesis that the Black Death originated in cosmic dust has been totally ignored in the standard historical works on the subject. But the Hoyle thesis has gained some surprising sympathy in scientific circles. In the 1980s Sir Francis Crick, the molecular biologist and Nobel laureate, who was codiscoverer of the structure of DNA, mounted arguments similar to Hoyle's, causing a momentary press sensation. In 1999 Paul Davies, a theoretical physicist and winner in 1995 of the prestigious Templeton Prize for Progress in Religion, published *The Fifth Miracle: The Search for the Origin and Meaning of Human Life*, which again inclined to Hoyle's belief in interplanetary transfer of organisms, which was again on this point ignored.

Any close scrutiny of the Black Death arouses concern that the conventional theory of diffusion by black rats is only part of the story. Even the addition of cattle-transmitted anthrax may not explain everything.

The medieval view that the origin of the plague involves floods and serpents arouses skepticism but cannot be ruled out. So how can we casually dismiss the periodic vertical transmission of plague and other infectious disease from outer space to earth?

It has been the fashion for academic medievalists, when they notice medieval writers placing the origin of the Black Death in rationally improbable events, wild, cataclysmic, and remotely violent happenings to insert such reports within a tradition of Christian apocalyptic literature. They argue that the vision of last things, especially in the Book of

Revelation, overdetermined these imaginative scenarios of earthquakes that liberated huge serpents that swam up rivers and spread disease.

But we cannot be sure that these hotblooded medieval explanations are simply derivative of artfully constructed biblical terror. They could have happened, just as the stories about King Arthur, Queen Guenevere, and Lancelot told around Welsh campfires in the early Middle Ages could have happened and then been dispersed to the far margins of literacy by the rationalizing state and church of the twelfth and thirteenth centuries. When an obscure administrator at the University of London, John Morris, took this line in *The Age of Arthur* (1973), the academics laughed at him—too readily. It is just possible that medieval writers who placed the origins of the Black Death in serpents dispensing plague as they swam up rivers were on to something.

Vertical transmission of disease from outer space in the tails of comets is our era's version of the extreme history of the Black Death. Although expressed in the context of astrophysics and endorsed by a handful of respectable scientists and science writers, the cosmic dust thesis strikes us very much like the medieval fixation on water serpents.

At a certain point, however—one we have not yet reached—extreme history begins to impinge on conventional historiography, and common consciousness has to acknowledge that things unique, horrendous, and otherwise inexplicable have in fact occurred.

CHAPTER NINE

HERITAGE OF THE
AFRICAN RIFTS

❧❦❧

I N THE 1960S AND EARLY 1970S paleontologists
made immensely important discoveries in the rift
(deep) valleys of East Africa, near the border of modern Kenya and Tanzania.

First, Mary Leakey discovered the footprints of earliest
man leading a small horse, the footprints embedded in hard-
ened clay. Then Grant Johannsen found part of the skeleton
of this earliest known human being, 2.5 million years old.
The humanoid was a four-foot-tall woman, probably black,
who was given the name Lucy because the tape recorder in
Johannsen's archeological campsite was playing the Beatles'
song "Lucy in the Sky with Diamonds" when this identifi-
ably bipedal (erect) human being was discovered.

It used to be debated whether earliest man appeared separately in various places on earth—Peking, Java, even Europe—or in one single place and spread out from there

during millions of years. The discovery of Lucy seemed to settle the argument in favor of the diffusionist theory. Various tests in the 1970s and 1980s proved she lived before any other known human.

Over millions of years, impelled by environmental changes and the gathering and hunting of food, the human species spread out from Africa over the rest of the world. Humanity reached Western Europe one hundred thousand years ago. A prime avenue of diffusion was the topographical funnel up the Nile from Kenya through the Sudan to the Nile delta and the Mediterranean. This is the great diffusionary chute of prehistory in the earliest times, a very long period before written records began in the fourth millennium B.C.

The Nile also remained the avenue in later times for the spread of antihumanoid epidemics, from medieval plague to modern AIDS, which also originated in East Africa. Lucy's homeland is the starting point of human society. It is also the starting point of diseases that have threatened the existence of humanity and in some cases still do so. Even the newest infectious disease, the West Nile virus that threatened New York City and Long Island in the summers of 1999 and 2000, originated in East Africa—in the West Nile district of Uganda.

Humanoids are cognates of the primate family, although we do not know exactly how this connecting evolution occurred. Lucy can be regarded as the black mother of us all and East Africa as the birthplace of the human race. The same area, however, was the incubus of infectious microbes that have threatened the existence of the human race over a long time. These diseases too came up the geographical chute from East Africa through the Sudan into the Nile delta and the Mediterranean. We know that for sure about medieval bubonic plague and modern AIDS, and we can speculate that was also the source and route of other biomedical scourges, particularly smallpox.

Why East Africa? Only in East Africa was there millions of years ago that necessary combination of climate, topography, and primates to start human evolution. The diseases that struck there—whatever their origins—were most effective there because elsewhere there was little or no life to attack.

There were pandemics far back in antiquity. The Hebrew Bible mentions them. They were Yawheh's instrument against the oppressors of the Chosen People, the Jews. God visited ten plagues on pharaoh and the Egyptians (still commemorated at Passover Seder by the spilling of ten drops of red Manischewitz wine) until bad pharaoh let Moses and his people go across the Red Sea toward the Promised Land.

The fearsome Philistines were also at one point struck down with an epidemic. The Assyrians had to give up their siege of Jerusalem because an epidemic broke out in their army as it lay encamped under the walls of the city . The latter biomedical event (if not the earlier ones), in the sixth century B.C., and recognized in the Bible, has some degree of historical plausibility.

Certainly there was a great plague in fourth-century B.C. Athens. It struck down the Athenian leader Pericles, and the deathly havoc it played on the Athenians contributed to their unexpected defeat in the Peloponnesian War with Sparta. Nobody knows what this pandemic was. Most historians guess it was smallpox, though it could have been bubonic plague. In any case it probably came up the mortality chute from East Africa into the Mediterranean and was brought along the trade route from Egypt, which the Athenians loved to visit.

There are plenty of learned people today who think that it was historically important that intellectually advanced and quasidemocratic Athens should have been defeated and demoralized by autocratic, militaristic, unintellectual Sparta. In his twelve-volume *A Study of History*, completed in the 1950s, Arnold J. Toynbee identified the Peloponnesian War as one of the great catastrophes of history, dooming western civilization almost as soon as it got started. But then Toynbee was a classical scholar and given to exaggeration on subjects like this. To the extent there is weight in Toynbee's view, disease played an important part by so weakening the Athenians that they lost the war.

Pandemics unquestionably shaped the course of world history with the biomedical waves of disaster that afflicted the Roman Empire and lay it open to attack and invasion from Germans and Mongolians in the late fourth and fifth centuries A.D. and Muslim Arabs in the seventh century.

The Roman Empire constituted the coalescence of the Mediterranean civilizations of antiquity under the political aegis of the Roman aristocracy and upper middle class. It stretched from Lebanon to Scotland, from Vienna to five hundred miles beyond Tunis into the North African Maghreb, from Paris to the southern border of Egypt and the Sudan. It was a highly literate and artistic and productive civilization, inventive in religion and law, urban-centered and peaceful among its ethnically diverse population, which had reached a level of fifty million by A.D. 250.

Then a series of plagues and pandemics assaulted the Roman world, and by the mid–seventh century A.D. the empire as a political and cultural unit was gone in Western Europe, replaced by violent, unstable, and predominantly illiterate barbarian kingdoms. All of the eastern and southern shores of the Mediterranean were by A.D. 700 ruled by Arab-speaking Muslim lords and even Anatolia (Asian Turkey) now was mostly under Arab rule. The Roman Empire had shrunk to Constantinople and a piece of the Balkans.

Since Edward Gibbon published the first volume of *The Decline and Fall of the Roman Empire* in 1776, historians have contemplated the causes of this great transformation in the Mediterranean and Western Europe. Gibbon thought it was due to overcommitment to otherworldly Christianity plus an excessive gigantism, the empire outgrowing its communications and transportation network. The Yale historian of the 1930s Michael Rostovtzeff attributed the downfall of the Roman Empire to class polarization and the takeover of power by illiterate masses, anticipating the fate of czarist Russia, from which Rostovtzeff had fled after the Bolsheviks took over.

Research since the 1950s has dramatically contributed to the conclusion that Rome's main problem was biomedical. From about A.D. 250 to A.D. 650 the Mediterranean world was assaulted by successive waves of pandemics that reduced the population by at least one-quarter, causing a manpower shortage in a society whose productivity was based on very little in the way of machinery and almost entirely on immediate plentiful human labor.

The result was far-reaching: a decline in food supplies and drastic reduction in industrial production and with it enfeeblement in exchange of goods through international trade. This caused shrinking of an already inadequate tax base, decreasing funds for bureaucracy and defense. A severe shortage of soldiers to defend the empire's immensely long frontiers occurred.

Gibbon was right. The conversion of the Mediterranean peoples to Christianity by A.D. 400 did engender a new and distracting otherworldly consciousness. The empire was too large and unwieldy and suffered as Gibbon said from "immoderate greatness." Class struggle and social and cultural polarization hurt the stability of imperial rule, as Rostovtzeff insisted. And there were some bad emperors along the way, who made countless political and military errors.

But it was the shrinkage of the population and the devastation and fear brought about by succeeding waves of epidemics that sank Rome and thus changed the course of history. Rome's greatest enemy came not from within but from biomedical plagues that the Romans could not possibly understand or combat.

The three pandemics were smallpox and gonorrhea from A.D. 250 to A.D. 450 and bubonic plague from 540 to 600. Where smallpox and gonorrhea came from is unknown. Some historians have guessed from the black hole in Central Asia. They may have just as well have come up the great mortality chute from East Africa. Certainly that is where the bubonic plague came from after A.D. 500.

Smallpox as a disease was officially declared extinct in 1978. It exists today only in the test tubes of a few research laboratories. But for fifteen hundred years it was one of the great menaces to human life. Because of the effectiveness of almost universal infant inoculation it has seemed of little importance, at least in the Western world, since the 1930s. But as late as 1800 smallpox was a terrible threat.

Before artificial immunization, populations that were not naturally immunized by surviving a previous outbreak of the disease were prey to almost incredible levels of mortality. Smallpox wiped out 9 million Native Americans in Mexico in the sixteenth century, the disease having been imported there by the naturally immunized Spanish conquerors. It had a terrible impact on the Roman world in the period 250 to 450, opening the way for the barbarian invasions.

Gonorrhea was the first sexually transmitted disease to emerge in the Mediterranean and Roman world. It too may have drifted in from Central Asia or from East Africa. It may have been spontaneously generated in the Roman Empire itself, which tolerated every kind of sexual intercourse known to man, including copulation with animals.

The appearance of gonorrhea had a cautionary impact on morals. Its prevalence in the fourth century allowed the Church fathers Ambrose and Augustine to condemn Roman promiscuity and prescribe puritan ethics. Only virgins were safe from sexual diseases, so they were elevated to saintly status in the church.

The epidemic waves caused terror and pessimism in a society with little true scientific knowledge. The ancient world had always relied heavily on faith healing, whether pagan or Christian. Now faith healing became the main recourse against omnivorous diseases devastating whole populations in the hitherto stable and comfortable Mediterranean world.

Like the first two pandemics that devastated the Roman world, smallpox and gonorrhea, the third wave, bubonic plague, had a terrible impact on the peoples it visited. The emperor of the mid–sixth century, Justinian I, the ruler in Constantinople who expended vast treasure building armies and fleets to invade and rescue North Africa and Italy from the German invaders, saw his military and political enterprise threatened by the devastating outbreak of bubonic plague in Constantinople and other great cities of the Mediterranean.

Justinian's successors were not able to hold back the Muslim armies from Saudi Arabia in the mid–seventh century. Roman defense that had held back the Arabs for millennia gave way in an empire impoverished and diminished by disease. While under frequent siege, Constantinople with its Church of the Holy Wisdom that Justinian had built held out until 1453, until it too was taken by the Ottoman Turks on their way to Bosnia.

The rise of political entities, legal systems, learning, urban living, and commercial productivity in medieval Europe from 800 to 1300 was made possible by a warming climate and absence of pandemics. There were some moments of effective leadership in state and church that helped the maximization of European growth in economy and learning. But it was the benign climatic and biomedical environment that was most responsible for the rise of European power and wealth, the clearing of land, the revival of cities, and above all the expansion of the population base fourfold from 900 to 1300. Europe was lucky, but in the long run it was also made more vulnerable by the long absence of pandemics. There was no capacity of natural immunization in these benign circumstances.

Thirteenth-century Western Europe extended from Iceland to Warsaw, from Oslo to Palermo. It was synonymous with Latin Christendom, with the area of devotion to the Roman Catholic Church, although within this spiritual umbrella there were plenty of critics of the papacy and clergy and at any given time at least 5 percent of the population had separated themselves from the Latin Church and set up their own religious ("heretical") communities.

The heartland, the richest and culturally most advanced parts of European civilization, lay in southern England, in northern France in the Seine-Loire valleys and the city of Paris, in southern France along the Rhone River, and in the Rhine Valley of Germany and the Low Countries at the great river's mouth, and in northern Italy from the Po Valley to Rome. This European heartland was much more thickly populated than the rest of Europe and was also the location of its great cities.

For reasons involving water and sewage and the capacity to encircle urban enclaves with high defensive walls, no city in the thirteenth century had more than 125,000 people. But the heartland areas were also dotted with small towns of 5,000 to 20,000 people and innumerable villages holding 500 to 2,000 souls.

In many respects the Europe of the thirteenth century, whose total population probably came close to the 50 million level of the Roman Empire at its peak, was a remarkably creative society and culture. Governmental agencies were instituting modern bureaucratic and legal systems. Well-endowed and heavily enrolled universities were the sites of brilliant theorizing about philosophy and theology as well as providing professional schools for training lawyers and teachers.

Magnificent Gothic churches were erected and the visual

arts in general—sculpture, painting, and colored-glass blow-ing—attained a level of ingenuity never to be surpassed. The vernacular national literatures of Europe were in the course of formation, and the texts invented by medieval poets and narrators are still intensely scrutinized today in all university literature departments for their subtle psychology and in-trinsic aesthetic value.

The Latin Europeans were afraid of neither the Muslim Arabs to the south nor the Greek Orthodox Slavs to the East and were pushing out their frontiers in those directions. It was a thriving, productive, and creative society and it was aston-ishingly peaceful—there was no major war fought in Europe between 1214 and 1296. In many ways the Europe of the thirteenth century resembled that of the nineteenth century.

But there was one important difference. Europe of the nineteenth century, at least in the second half, invested heavily in scientific research and laid the basis thereby for first The New Physics of the early decades of the twentieth century and then the biomedical revolution after 1940. Thirteenth-century Europe, aside from research on optics, which led to introduction of eyeglasses, and improvement in mechanical clocks, made no scientific progress.

Short of algebra before the sixteenth century, prohibited by Church restrictions and popular attitudes from undertak-ing dissection of the body so that even the blood-pumping function of the heart was not known, and lacking professor-ships in natural science in the universities, the intellectual realm attained wonders in theological and philosophical speculation but in understanding nature barely advanced beyond classical antiquity.

Microscopes and telescopes did not appear before 1600. The favorite science textbooks of the thirteenth century were those of Aristotle, written in the fourth century b.c. Aristotle was an overachieving genius but he was fundamentally in the wrong about scientific essentials and until Aristotle was discredited around 1600 Europe was stuck in an intellectual cul-de-sac.

This Latin Christendom of the thirteenth century was an immensely creative but amazingly one-sided culture. It fatally did not apply its resources to scientific research, whether physics or biology. It had some knowledge of chemistry but wasted it on alchemy, trying to turn base metals into gold. It had a little knowledge of astronomy but wasted it on a rage for astrology and fortune-telling.

Europe was weakest in biomedical areas. Except for a few eccentrics like Roger Bacon, the Oxford Franciscan, it was an arrogant, heedless culture that could build a magnificent church and develop a new legal system as well as any culture, including our own, that has ever existed. But it had no understanding of disease, neither its nature nor its cure, and it was extremely vulnerable to epidemics. Essentially it had only nonbiomedical responses to devastation of a breakdown in societal health—pray very hard, quarantine the sick, run away, or find a scapegoat to blame for the terror.

The rarest attribute in any society and culture, when things are generally going well, and peace and prosperity reign, and bellies are full of good food, and the sun shines and the rain falls appropriately, is to notice certain cracks in the edifice, some defects and problems, which if not attended to could in time undermine the happy ambience and bring on distress and terror.

We are today not as good at this self-scrutiny as we ought to be, but thirteenth-century Europe was far worse.

This was partly due to unbroken political and economic progress since about 950 and partly due to the Christian outlook embraced by the intellectuals and power-brokers: the almost universal belief that Latin Christendom was on direct route to ever happier days culminating in the Second Coming of Christ, which was only postponed until all peoples of the earth joined the Church.

What pessimists there were in the thirteenth century mainly expressed their doubts about constant felicity in an apocalyptic format called Joachimism, after a south Italian abbot and preacher who died in 1200, Joachim of Flora. The Joachimists claimed that the world was entering an era in which Satan would sit on the throne of Peter, no less, in Rome, and there would be darkness and terror before the Second Coming. This pessimistic view developed from the biblical Book of Revelation and possibly the Jewish mystical Kabbalah. It was condemned by the holders of power and by most intellectuals and combated by the papacy and was confined to a small minority. Since it was an entirely religious projection, it would not have been much help in analyzing the defects in Europe's material infrastructure and remedying them.

Thirteenth-century Europe was suffering from a classic Malthusian situation (named after the gloomy nineteenth-century economist). Its booming population, made possible by a long period of unusually warm weather and no epidemics, producing an adequate food supply and with it improved nutrition and longer lives, was beginning to strain at the limits of the food supply and the available space for increasing agricultural production to keep pace with the escalating population. Every premodern society has at one time or another experienced this Malthusian crunch.

Medieval Europe could do nothing about it except to turn its slash and burn methods of clearing forests against the steadily shrinking wooded areas, or to bring under the plow hilly and rocky countryside not suitable for cereal crops and hitherto used exclusively for pasturage. Real estate prices rocketed upward along with the population boom, and the price of grain and meat, the staple of the food chain, by the mid–thirteenth century was becoming an obstacle to the comfortable life of the peasantry and urban working class.

In premodern societies there was no agricultural chemistry to improve crop yields. The only ways out of the Malthusian maelstrom were a physical disaster, in the form of bad weather and crop failure generating widespread mortality from famine, and man-made terror in the form of warfare and massacre of the civilian population. Europe in the first half of the fourteenth century experienced both of these cruel correctives, enervating health and engendering a general feeling of pessimism and lassitude toward secular problems, and thereby providing a particularly adverse context for the outbreak of the Black Death.

Around 1290 Europe's good times ended, and historians debate when they returned—some say around 1500 but more recently it is claimed not until 1700. In the same quarter of a century Europe was hit by the end of the long peace, and by bad agricultural weather. It grew colder and wetter and for two summers in the second decade of the fourteenth century there was widespread crop failure in Western Europe and a great famine that began to reduce the population. Historians wonder whether this deteriorating weather was a natural cyclic change that occurs every two or three centuries in the northern hemisphere or if it was brought about by a specific natural disaster—volcanic eruptions in the East Indies that within two years had blocked out the sun in Europe for months at a time with an ashen cover in the atmosphere. Possibly both events occurred simultaneously.

Compounding the climatic changes were the man-made disasters of the very long and ruinous war between Europe's two powerful monarchies, those of France and England. While conventionally known as the Hundred Years War that began around 1340 it was more like a 150-year war that commenced in 1296. France and England plunged into a century and a half of imperialistic conflict over the textile cities of Flanders (Belgium) and wine-growing regions of Gascony (Bordeaux). There were long truces and interruptions, but the general effect was wasting of the French countryside in the western third of that country.

The war also generated organized rural crime in England headed by the heavily armed and well-trained demobilized soldiery who attacked England's civilian population when they were not sacking the French villages and towns. Warfare erupted in other places—the English and Scottish border, Spain, and parts of Germany and Sicily.

The good times were over. Hunger and violence stalked Europe on the eve of the Black Death of the late 1340s. But the new cataclysm immensely exceeded the already prevailing disasters. The deterioration of the food supply due to bad weather and war had weakened the resistance of Europeans to infectious diseases.

CHAPTER TEN

Aftermath

BEAUTY CAME IN THE late morning under the warm sun and reached fulfillment in the afterglow of the late afternoon. Then darkness fell and the cold bleak wind rustled the landscape under a chilly moon.

It is impossible to get this picture of the European Middle Ages, or some similar metaphor drawn from nature and climate, as in Johan Huizinga's *The Autumn of the Middle Ages* (1919), out of our heads. It is ingrained in our stereotype of the medieval historical arc—the early, high, and late Middle Ages. In the metaphorical construct, of course, the Black Death stands as the moon of cold darkness, or the onset of harsh bared-tree autumn passing into winter.

The Dance of Death, macabre skeletons rising en masse from murky graveyards, was a favorite motif of art and literature in the century following the Black Death of the 1340s. Huizinga never doubted that artistic motifs were a screen on which were projected the ideological anxieties of a society.

For him the Dance of Death exhibits pessimism, lassitude, and loss of confidence on the part of the courtly culture of the late medieval aristocracy unable to confront and control the realities of life.

We can at least go this far with the incomparable Huizinga (who wrote *Autumn* in a few months after his department head reminded him it was publish or perish time). The chaotic morbidity that the era of the Black Death featured would give artists the idea for the Dance of Death. It was, after all, not a very subtle motif.

And then what followed? How did Western society advance from the meanness and turmoil of the Black Death era, when bodies piled high in the streets, and mass graves were thinly covered by hastily piled earth, leading to the horrid stench of the simultaneous decay of thousands of bodies? What if anything has all this death and destruction to do with the onset of the Renaissance, the golden glow of neoclassical courtly and artistic creation at the end of the fifteenth century?

This question has challenged the wit and imagination of historians, inspiring answers all the way from nothing to everything. The Black Death came and went and left barely a trace of anything, said David Knowles in 1962. It made possible the Renaissance and proto-modern world, by breaking up the old culture, according to David Herlihy in 1995. The Black Death was the trauma that liberated the new.

It can be readily seen that the Black Death accelerated

the decline of serfdom and the rise of a prosperous class of peasants, called yeomen, in the fifteenth century. With "grain rotting in the fields" at the summer harvest of 1349, because of labor shortage, the peasants could press for higher wages and further elimination of servile dues and restrictions. The more entrepreneurial landlords were eventually prepared to give in to peasant demands. The improvement in the living standard of many peasant families is demonstrated by the shift from earthenware to metal cooking pots that archeologists have discovered.

The Black Death was good for the surviving women. Among the gentry, dowagers flourished. Among working-class families both in country and town, women in the late fourteenth and fifteenth centuries took a prominent role in productivity, giving them more of an air of independence. The beer- and ale-brewing industry was largely women's work by 1450. The growth of a domestic wool-weaving industry allowed working-class women to become industrial craftsmen in the textile industry. The graphic picture of farm women churning butter in their kitchens that George Eliot gave us in *Adam Bede* (set in the 1790s) was certainly occurring by 1400.

As in all primarily rural societies during times of economic upheaval, there was a flocking of "misdoers"—criminals, beggars, and prostitutes—to London from the countryside. It was like many parts of sub-Saharan Africa and Latin America today.

The biomedical devastation had a strange and complex impact on the Church. It may have reinforced a trend away from optimism to pessimism, from a God who could be partly encapsulated in reason and was a mighty comfort and fortress, to one whose majesty and planning and rationale were impenetrable, although that pessimistic inclination was already rising in intellectual circles thirty years before the Great Pestilence.

The century after the Black Death was marked—in England, France, the Low Countries, and Germany—by what may be called the privatization of medieval Christianity. This took both organizational and spiritual forms. Organizationally there was a rush by the affluent upper middle class to found chantries, private chapels supported by one family or a small group of families. The great lords and millionaire gentry and merchants had always had private chapels. Along with the capability of having three hundred people for dinner in your household, it was the signal conspicuous consumption of great wealth.

In the more plebeian chantries, the rising middle class imitated their betters. Even the workers organized into craft guilds got into the act. The labor corporations also became confraternities that sustained private chapels and provided burial benefits to their members.

Spiritually and intellectually, the century after the Black Death in England and elsewhere in northern Europe was marked by the rise of intense personal mysticism and separately by a privatist kind of bourgeois behavior in elaborate spiritual exercises.

The most remarkable mystical text was the anonymous *The Cloud of Unknowing* of late-fourteenth-century England. A reasonable guess would attribute authorship to some Carthusian monk. What is remarkable is *The Cloud* comes close to the concept of nirvana in Oriental religion. This is negative rather than positive mysticism, which was heretofore the main-line medieval form. *The Cloud* offers not a reaching out to God in the twelfth-century Neoplatonic manner, which was now assumed to be impossible, but an emptying out of all intellect, imagination, and feeling from consciousness; a nirvanic condition of total negativity and depersonalization that allows the inrushing of God's love and majesty into individual souls.

The other shift in late medieval religious sensibility was compulsive focus on the body of Christ. This inspired the elaboration of the Corpus Christi festivals and procession in the late medieval town and countryside. It drove the new fashion of taking the sacramental wafer as often as possible, instead of the old prescribed minimum communion of once a year. Eating Jesus in the Mass became the self-help mode of late medieval Christianity.

It is, however, impossible to answer the question of whether these departures in the institutional and emotional sides of medieval Christianity were a reaction to the pessimism, loneliness, and despair of the plague years.

Probably these changes would have happened anyway. They were built into late medieval social trends, the expression of the middle class, and attitudes of the yeoman class. They are examples of "asceticism in the world" that Max Weber associated with the Protestant Reformation of the early sixteenth century but that antedated the Reformation by more than a century.

The Black Death provided an activating psychological context for privatization of late medieval religions. It did not create it.

Its direct impact on the Church was more in the way of affecting personnel. At least 40 percent of the parish clergy, equal to the mortality rate among the peasants and workers they ministered to, were in the late 1340s carried off. Some cathedrals' chapters were close to being wiped out, and many abbeys were similarly devastated, even though these were privileged precincts.

This meant a shortage of personnel to maintain staffing levels and gravely threatened institutional stability and continuity. The solution was to recruit and appoint much younger men, after petitioning the archbishop and pope to relax age requirements set down in canon law during the population explosion of the late twelfth and thirteenth centuries.

During and immediately after the Black Death, priests were ordained at twenty rather than twenty-five. Monastic vows could be administered to adolescents at age fifteen rather than twenty. Priests took over parish churches at age twenty instead of twenty-five. It was a younger, much younger Church that came suddenly into being, and one now staffed heavily with undereducated and inexperienced people.

University graduates previously redundant in their numbers and underemployed, like humanities Ph.D.'s today, were strongly in demand in the post-Death years. They disturbed the surviving older generation of church officials by bargaining for higher pay and greater privileges. Almost immediately the excess employment pool of university graduates was drained to zero.

This had the unanticipated effect of driving the spread of the Lollards, the feared radical heretics whose founders came out of Oxford seminars, especially John Wycliffe's, to attack church leadership, and ecclesiastical morality, and to question even the efficacy of the Sacrament of the Mass. The Lollards also aroused fear and anger in the established Church by allowing women to preach in their communities. The Lollards could point to the inadequate quality and quantity of ecclesiastical staffing, the ignorance of the parish priests, and the greed and selfishness of the monks. These complaints were echoed in the writings of the contemporary poets.

Now these university graduates preaching Lollardy in country and town could advertise their superior qualifications as Christ's ministers to those often displayed by downscale newly ordained clergy and monks.

The Lollards with their superior learning and unwavering piety—in contrast to the now-tattered norm of ecclesiastical staffing—were able to set up counter-churches. When their teachings were too democratic and revolutionary for irate royal officers and courtiers, the Lollards withdrew to distant districts among the northern granges and ranches. They were still there in the 1530s when Henry VIII broke with Rome so he could make an honest married woman of his pregnant mistress.

Changes in artistic style as well as spirituality have commonly been attributed to the Black Death.

In 1951 the art historian Millard Meiss saw this trend in north Italian painting after the Black Death—a more religious, less humanistic style. Meiss postulated a throwback to the abstractions of twelfth-century art and away from the naturalistic humanism of Giotto in 1300. This indeed may have happened in Florentine and Sienese art in the two or three decades after the Black Death. But not necessarily for intellectual reasons. The small number of progressive ateliers following Giotto's innovative naturalistic humanism were possibly wiped out by the plague. The conservative ateliers, the majority, remained sufficiently active and followed the old twelfth-century abstractionist style.

By the time Meiss wrote his classic work in 1951, the connection between higher culture and social change that had been propounded by the Marxists of the Frankfurt School in the 1930s, Theodor Adorno and Walter Benjamin, and the Italian communist theorist Antonio Gramsci, had come to be widely accepted and applied in academic circles.

It was the theory of structure: An economically shaped society determines the superstructures of art, literature, philosophy, and science. The European Marxists postulated the possibility of a small degree of autonomy for the superstructure only after it had been established on its material and social base, and Meiss's book is an example of this flawed theory.

What we have come to know about cultural change in the twentieth century from experience and observation raises doubts about the structure/superstructure paradigm. There is an integral aesthetic, psychological, highly personal quality to innovative art. Aspects of craftsmanship also enter into the making of a style. Great art and literature elude the rules of sociology.

In England there was a parallel increased austerity in architectural style, which can be attributed to the Black Death—a shift from the Decorated version of French Gothic, which featured elaborate sculptures and glass, to a more spare style called Perpendicular, with sharper profiles of buildings and corners, less opulent, rounded, and effete than Decorated. It may be due to more mundane causes—running out of the supply of French masons and sculptors because of much lower immigration into England during the Hundred Years War, as well as a modicum of nationalist puritanism opposed to French decadence during that era. The cause may have been economic—less capital to spend on decoration because of heavy war taxation and reduction of estate incomes because of labor shortage and higher peasants' wages.

Can we say something more firmly about the connection between the Black Death and the Italian fifteenth-century

Renaissance—its humanistic learning, its more accessible literature, its more naturalistic depiction of the human face and body in painting and sculpture?

The Black Death hit many of the great cities of northern Italy very hard. Giovanni Boccaccio's *Decameron* uses the realistic flight of the Florentine rich and beautiful to the—they hoped—safer countryside and the glitterati's diversion there by telling and listening to sexually charged stories as a framing device for his volume of romantic and sarcastic tales.

Since the *Decameron* is considered one of the launchpads of the Renaissance, and since the historian William Bowsky in the 1960s discovered in the archives of nearby Siena an affluent merchant whose five children died in the plague, the question has arisen: Did biomedical trauma then somehow trigger the Italian Renaissance?

Perhaps it had some impact on consciousness, but the trend developed in the fifteenth century of Renaissance culture was in the direction not of more soul searching but of a more earthly and secular experience. Perhaps the Black Death weakened faith in traditional medieval Catholic spirituality and set off a quest for a deeper naturalistic understanding of human psychology and behavior and the expression of a more personal sensibility.

This possible tangential result of biomedical devastation scarcely diminishes what we know to have been the main driving forces in Italian Renaissance culture—a very rich and politically powerful high bourgeoisie trying to establish its own identity relative to aristocratic and ecclesiastical sensibility by recalling and identifying with the frame of mind, taste, and behavioral patterns of the ruling class of ancient Rome portrayed in its abundant literature and surviving art.

Somewhere in the common unconscious there may have sounded the tocsin that the Black Death was the symbolic psychological decline of the old world of the European Middle Ages. It was time to get on with the new and modern, given added legitimacy by renewed classicism. But like all of cultural history, this is just an ephemeral and unprovable observation.

In 1919 Johan Huizinga was sure that the late medieval Western European culture was characterized by a noticeably increased scrutiny of and reflection on death. This enhanced consciousness of morbidity was held to have been expressed in the art and literature of what he called the autumn—or waning—of the Middle Ages.

Since Huizinga two remarkable books have explored this theme. Jean Delemeau of the French school of social and cultural history in 1991 perceived the sharp rise of a sense of sin and fear after 1300. He saw "the emergence of a Western guilt culture," which endured until the eighteenth century. In 1996, Paul Binski, the Cambridge University art historian, highlighted the sense of the macabre in late medieval culture. Binski tells us that fourteenth-century penitential expression stresses "that the individual is subjected to the inevitability of the larger and ineluctable order of mortality."

More elaborate and sometimes eccentric funeral arrangements represent for Binski "a masochistic reversed decorum by which courtly aristocratic norms of behavior could be exposed in all their elegant vacuity." Binski goes on to speak of the rise of Delemeau's guilt culture, centered in "the macabre" and—suddenly remembering the Renaissance—he boldly claims that "secular individualism and guilt culture represent two sides of the same coin," a familiar Freudian ambiguity.

Death imagery in late medieval literature and art, Binski concludes, "is fundamental to the ways a culture thinks and to the way it allegorizes meaning." But he discounts that the advent of the macabre "was owed to a specific exogenous, external, socially contextual cause like the Black Death. This period of the macabre . . . can only be understood by seeing it as an internal development of medieval visual culture itself." He is scathing in his critique of Millard Meiss's 1951 social reductionism, and acutely suggests that one of the key paintings used by Meiss to show the reversion to spiritual abstraction (or "transcendentalism") after the Black Death may have actually been executed before the Great Plague— the perils of art history.

The Huizinga-invented, Delemeau- and Binski-articulated increase in death consciousness in the fourteenth century may very well be the work of retrospective imagination that historians of all kinds frequently exercise. There is no change in the intellectual quality of death theology between

1200 and 1400. What was visibly different was a quantitative increase in expression of this morbidity in art and literature. This was due to increased artistic and literary productivity and the chance survival of texts and artworks.

If, however, Huizinga, Delemeau, and Binski are right and there was a qualitative, intellectual rise of a death culture, that development may have significantly affected the manner of social response to the Black Death. A culture so extremely focused on mortality, sin, and macabre, posthumous punishment would be in a poor state of mind to take practical steps to counter the Great Pestilence. A society moving toward secular individualism and nonhierarchical communal action might have been expected to take more affirmative action to counter the plague through improvements in medical science and social organization. But there were no such advancements. If European culture indeed entered a new era of death consciousness, funeral ritualization, extravagant guilt, and macabre imaging after 1300, this cultural ambiance facilitated the incapacity for human responses to the Black Death. The Black Death did not create the Dance of Death; in other words, the causality ran the other way.

The Black Death's impact on the course of the Hundred Years War is much more clear than its tenuous relationship to the Italian Renaissance. At a time when infantry had become central to English tactics, the cost of soldiers drawn from the peasantry increased because of the drastic population decline from the plague. This meant a sharp escalation in military personnel expenses for the English monarchy and made the ultimate English victory much more costly and difficult.

Without the Black Death and the 40 percent reduction in the peasant population from whom the infantry were drawn, it is just possible that the Plantagenets would have made themselves kings of France. After the Pestilence and demographic crash that was a task too challenging even for the relatively advanced taxation system of the medieval English state.

In the end what happened to the Plantagenets' Anglo-French empire was very similar to the fate of the Roman Empire. Both were brought low by a biomedical devastation that caused a sharp fall in the size of the working and military population. Both were grand edifices undone by the specter of infectious disease and pandemics. There is a lesson for the American empire today in that situation.

The cataclysm of the Black Death weakened the foundations of medieval kingship. In the words of Shakespeare in *Richard II*, still in some ways the best account of that pathetic and unstable figure, it threatened to "wash the balm from off an anointed king."

Medieval kingship was built on foundations that ran back centuries in time: divine grace attributed to kingship's office by the Church; Roman imperial and legal authority; war leadership and national feeling derived from Germanic sources. Edward III and his heir Edward the Black Prince

had shown that they could win battles in France but had
nothing with which to fight the Pestilence. They ran to their
most remote country estates to save themselves and left
nothing for society to do except pray.

The Black Prince, struck down by malaria from his ill-
fated Spanish invasion, predeceased his aged, venereal-dis-
ease-ridden father. It was the Black Prince's wan, nervous,
uneasy, high-strung son who succeeded the old man in 1377.

Richard II's twenty-year reign ended in the ignominy of
his forced abdication in 1399, as a denouement of a small
revolution, concocted by Richard's first cousin Henry of
Lancaster, John of Gaunt's son. Previously Richard had ban-
ished his rival Henry Bolingbroke from the kingdom, then
mischievously—in unchivalric manner—deprived Henry of
access to the wealth of his Lancastrian lands. Henry invaded
England from France with a small mercenary army while
Richard was fighting an unsuccessful war in dismal Ireland
and the nobility, the churchmen, and many of the king's
own courtiers rallied to Lancaster.

By the time Richard made it back to England he was utterly powerless. Henry forced his abdication, and then to
make sure of his cousin's elimination, called a Parliament
that declared Richard a tyrant, a Romanist enemy of the
common law, and deposed him. Parliament made Henry
king, filling the vacant throne with the artful and seemingly
benign usurper. Richard was taken off to a rural castle and
was likely starved to death.

Contemporary writers and modern historians have suggested a variety of scenarios to account for the downfall of Richard II.

He was, said William Stubbs in 1870, a tyrant who claimed with Roman pretension that law was in his own mouth and breast and had to be done away with because of such unconstitutional ideas and behavior. Richard had to be removed so that the proto-modern "Lancastrian constitution" could be activated.

A second explanation, offered by recent biographers Nigel Saul and Michael Bennett: Richard, like John Lackland and Edward II, lacked the political temperament to be king. A man of learning, piety, and fine aesthetic taste, Richard interacted awkwardly and impatiently with the great nobility. They experienced his mood swings and ferocious anger and finally got tired of him and afraid of his next moves. Lancaster's bonhomie seemed much more accommodating. This interpretation resembles Shakespeare's play, which Saul strongly admired.

A third explanation, suggested at the time: Richard was homosexual. There were no children from his diplomatic marriage to Ann of Bohemia. Like Edward II, he built up an inner set of gay courtiers. This inflamed aristocratic mistrust with regard to the rationale behind royal patronage and antagonized the bishops, who had long departed from the early medieval church's tolerance of homoerotic behavior.

Richard tried hard. He brooded over what to do, how to

build up his image, how to relate to the nobility and other people. Feverishly dissatisfied with one course of action and one set of friends, he swung in jagged fashion to contradicting policies and groups. He knew that peace with France was necessary for the welfare of his kingdom but characteristically was never able to negotiate a durable peace settlement. Over the last three years of his reign he became vengeful, paranoid, and increasingly erratic and unpredictable. Nobility and Parliament reluctantly concluded that Henry of Lancaster was a better or at least less dangerous man.

What really haunted Richard was the cataclysm of the Black Death, of the shadow of extermination falling upon society while the proud Plantagenet monarchy stood muted or ran away to escape into the safer countryside. Richard tried to restore royal leadership through activist control over events in the manner of Henry II, Edward I, and his grandfather Edward III. But he did not know what to do except to act in an arbitrary and noisome manner.

Shakespeare was right that Richard II was a tragic figure. His tragedy was rooted in a compulsive desire to restore the grand and popular Plantagenet monarchy that inexplicable biomedical calamity had assaulted.

Henry of Lancaster as king (1399–1413) was not much more successful. He exhibited not hyperactivity but lassitude and indecisiveness. Like Richard II, Henry IV was frustrated and stymied by a postplague condition of radically diminished human resources.

Henry IV's son Henry V (1413–1422) revived the Hundred Years War and archaically regained military luster and national pride for the monarchy by his lucky win at Agincourt in 1415. Seven years later Henry V died prematurely as he seemed about to become king of France. He left a child king who grew up into a mentally deficient and unstable adult. Henry VI (1422–61) brought down the proud Lancastrian family, losing their French empire and then the throne itself to his ruthless Yorkist relatives.

It was a long way down from the sunny day when Princess Joan in 1348 had sailed from Portsmouth to Bordeaux to journey to Castile and there build through marriage the Plantagenet empire in Spain. When Joan landed on the quay in Bordeaux harbor the rise of the Plantagenets to European dominance seemed unstoppable. The bacillus from a flea-ridden rat or consumption of beef from a sick cow that killed her altered the course of European political development for the next hundred years.

The biomedical catastrophe took away charisma from kings, eroded popular support for their veneration and self-esteem as God's anointed and as war leaders and money providers. It drove a sensitive, intelligent monarch like Richard II toward anguished behavior and antisocial, politically imprudent policy that led the nobility headed by his cousin to bring him down and kill him.

In 1400, the year that Richard II perished in bleak Pontefract Castle, England's greatest medieval poet, Geoffrey

Chaucer, also died. The king and poet would have known each other slightly, Chaucer being a leading court poet, an assiduous servant of the Lancastrian family, a diplomat, and a collector of customs in the port of London. Because of Chaucer's very close connection to John of Gaunt and the Lancastrians, the king would have kept a wary eye on him. He did not offer Chaucer further royal patronage beyond what Duke John had proffered.

Chaucer was the son of a middle-class wine merchant. He did not attend university and was largely self-educated, and profoundly so, in Latin classics and French and Italian Romantic and Humanistic literature. We know little of Chaucer's life itself. Early in his career he was accused of rape by a woman of substantial family, and the case was settled out of court. Donald Howard, Chaucer's biographer, believes the poet's wife was possibly one of John of Gaunt's mistresses; likely her sister was. On a diplomatic trip to northern Italy it is possible that Chaucer sought out and met the Florentine poet and classical scholar Petrarch. Other than that, we know that Chaucer spent plenty of time on missions to France during the Hundred Years War; the rest is obscure.

Chaucer's work proves that he was prolific in adapting French and Italian literature, including Boccaccio's *Decameron*, pillaging its more lusty scenes. The famous Princeton University critic D. W. Robertson, Jr., argued vehemently in the 1950s that Chaucer was deeply committed to Augustinian theology. After a fashion he may have been, but Chaucer was not a didactic theologian. His Christianity was implicit in his literary work and his religious views were conventional, including fervent anti-Semitism, even though there were almost no Jews in England at this time.

Chaucer was different from Richard II in his reaction to the Black Death. Chaucer was not torn by anxiety to countervail the pandemic's consequences, as the king was. Unlike his contemporary William Langland in *Piers Plowman*, Chaucer ignores the plague.

The Canterbury Tales is a work of what today would be called journalism, which accepts all the horrors, ironies, and complexities of social life and passes no ostentatious judgment on them. Chaucer the journalist gives us human interest stories of middle-class people, women as well as men, and persuasive three-dimensional profiles. He seeks to intrigue and amuse. He has come along in a world that has been crippled by a disaster but that struggles successfully to resume the rhythm of its daily life, its occupations and little pieties, its sexual tensions, its prejudices and assumptions. A social healing is occurring but Chaucer does not reflect on this process: He illustrates its happenings.

Knowing About the Black Death

A Critical Bibliography

HISTORICAL CONTEXTS

These are indispensable introductions to the political and social context in which the Black Death in England occurred: Maurice H. Keen, *English Society in the Later Middle Ages* (London: Penguin, 1990); Stephen H. Rigby, *English Society in the Later Middle Ages: Class, Status, and Gender* (New York: St. Martin's, 1995); Michael Prestwich, *The Three Edwards* (London: Weidenfeld and Nicholson, 1980); W. Mark Ormrod, *The Reign of Edward III* (New Haven: Yale University Press, 1990); John Hatcher, *Plague, Population and the English Economy* (London: Macmillan, 1977); Christopher Dyer, *Standards of Living in the Late Middle Ages* (New York: Cambridge University Press, 1989); Paul Strohm, *Social Chaucer* (Cambridge, Mass.: Harvard University Press, 1989); Clifford J. Rogers, *The Wars of Edward III* (Rochester, N.Y.: Boydell, 1999); Rodney H. Hilton, *Bondmen Made Free* (New York: Methuen, 1977); Barbara W. Tuchman, *A Distant Mirror: The Calamitous Fourteenth Century* (New York: Knopf, 1978); Jonathan Sumption, *The Hundred Years War*, 2 vols. (Philadelphia: University of Pennsylvania Press, 1991, 1999); Nigel Saul, *Richard II* (New Haven: Yale University Press, 1997); and Michael Bennett, *Richard II and the Revolution of 1399* (Gloucestershire:

Sutton, 1999). The works by Prestwich, Rigby, and volume one by Sumption are particularly illuminating.

On literary, intellectual, and religious history two works originally published in 1933 remain important: G. R. Owst, *Literature and Preaching in Medieval England* (New York: Barnes and Noble, 1966); Karl Young, *The Drama of the Medieval Church* (Oxford: Clarendon Press, 1967). In addition are three recent substantial works: Gordon Leff, *Bradwardine and the Pelagians* (London: Cambridge University Press, 1984); Edward Grant, *The Foundation of Modern Science in the Middle Ages* (London: Cambridge University Press, 1996); Norman Kretzmann et al., eds., *The Cambridge History of Later Medieval Philosophy* (Cambridge: Cambridge University Press, 1982). Donald R. Howard, *Chaucer: His Life, His Works, His World* (New York: Dutton, 1987), is the best book on the subject and brilliantly written; Paul Binski, *Medieval Death* (Ithaca, N.Y.: Cornell University Press, 1996), is careful, learned, and insightful; Jean Delemeau, *Sin and Fear* (New York, 1991), is weird and verbose but interesting; Ann Hudson, ed., *Selections from English Wycliffite Writings* (Toronto: University of Toronto Press, 1997), has an extremely valuable introduction and notes; Caroline Walker Bynum and Paul Freedman, eds., *Last Things: Death and the Apocalypse in the Middle Ages* (Philadelphia: University of Pennsylvania Press, 2000), is also useful, particularly the paper by Laura A. Smoller.

Works that argue for a strong impact of the Black Death on art and literature are Millard Meiss, *Painting in Florence and Siena After the Black Death* (Princeton: Princeton University Press, 1957, 1964); Daniel Williman, ed., *The Black Death: The Impact of the Fourteenth Century Plague* (Binghamton: N.Y.: Center for Medieval and Renaissance Studies, 1982), especially the major papers by Aldo S. Bernardo and Robert E. Lerner. Meiss famously argued that after the 1340s calamities there was a temporary reversion in north Italian art back to the quasi-abstraction of the twelfth century and away from the humanistic naturalism of Giotto. We do not know what would have happened in art and literature without the Black Death, but my counterfactual speculation is that the cultural development of the late fourteenth century would have been much the same without the plague.

On monastic life after the Black Death an instant classic was Barbara F. Harvey, *Living and Dying in the Middle Ages* (London: Oxford University Press, 1993), drawing heavily from the detailed records of Westminster Abbey. Remember how late medieval poets and satirists portrayed monks as fat and gluttonous and this hostile view was deprecated by generations of historians? It turns out that the poets and satirists were right.

On the Black Death in Italy, see William M Bowsky, *A Medieval Italian Commune: Siena Under the Nine 1287–1355* (Berkeley: University of California Press, 1981); John Henderson, "The Black Death in Florence," in S. Bassett, ed., *Death in Towns* (Leicester: University Press, 1992); Samuel K. Cohn, Jr., *The Cult of Remembrance and the Black Death: Six Renaissance Cities in Central Italy* (Baltimore: The Johns Hopkins University Press, 1992).

Ultimately, definitive knowledge of the biomedical and social components of the Black Death will not be achieved unless two vast research projects are systematically undertaken and come close to completion.

A. The historical human genetic map, going back 144,000 years, is nearing completion. Use of DNA studies is needed to determine the physiological and biochemical ingredients of the pandemic of the mid–fourteenth century, involving genetic analysis of human fossils and preferably cell tissue. Cell tissue is most likely to be found in bodies dug out of Arctic permafrost. From work done so far, an inherited genetic structure derived from the Black Death appears to have provided immunity from HIV/AIDS today.

B. A thorough combing has to be undertaken of the vast judicial, manorial, and urban records of fourteenth-century England to establish the trends in demographic, social, economic, and political change. Zvi Razi's studies of the manorial court rolls in the 1980s and 1990s for population statistics (especially *Life, Marriage, and Death in a Medieval Parish*, Cambridge University Press, 1980) showed what could be achieved by such microcosmic work. The vast English judicial records, royal and local, alone constitute enough labor for a generation of scholars: There is little incentive and reward system to undertake this gargantuan project. Learned foundations are no longer interested in this kind of long-term project.

GENERAL WORKS ON THE BLACK DEATH

Phillip Ziegler, *The Black Death* (London: Sutton, 1997, originally published in 1965). The text did not change in subsequent editions and printings; this edition added pictures. Highly readable and out of date.

Jean-Noel Biraben, *Les Hommes et la Peste*, 2 vols. (The Hague: Mouton, 1975), is a characteristic product of the French Annales school—verbose, unfocused, and now out of date. But information in the book on the Black Death in France, particularly Bordeaux, remains valuable.

Robert Gottfried, *The Black Death: Natural and Human Disaster in Medieval Europe* (New York: The Free Press, 1983). Similar to Zeigler's book but better informed; a useful synthesis. But Graham Twigg's biomedical study, appearing a year later, gave new perspectives. Excellent bibliography. The text could have been much better.

David Herlihy, *The Black Death and the Transformation of the West*, ed. and with an introduction by Samuel K. Cohn, Jr. (Cambridge, Mass.: Harvard University Press, 1997). The publication date is deceptive. The book's three chapters were delivered as public lectures at the University of Maine in 1985 and published with very little change by Cohn, Herlihy's student, at the urging of Herlihy's devoted widow. The first chapter on "Historical Epidemiology and the Medical Problems" is worth close reading, although Herlihy's dismissal of Twiggs's book, published the year before his 1985 lectures, significantly derogates from the value of Herlihy's argument.. The last two chapters of Herlihy's book are puff pieces of speculative nature, unusual for Herlihy, and are of modest value. Cohn's astute introduction is excellent.

Colin Platt, *King Death: The Black Death and Its Aftermath in Late-Medieval England* (Toronto: University Press, 1997; first published 1996). As in all of Platt's writings, the archeological knowledge is impressive and the selection of pictures very good. The book is mainly about the social and economic impact of the Black Death and relies on recent research, which is, however, fragmentary and superficial in relation to the totality of unpublished (and unstudied) manuscript sources.

JEWS

For a short account from a secular point of view and based on all the recent scholarship, see Norman F. Cantor, *The Sacred Chain: A History of the Jews* (London: Fontanta, 1996). Major detailed works are Salo W. Baron, *A Social and Religious History of the Jews*, 2nd. ed., vols. 8–10 (New York: Columbia University Press, 1971–73), immensely learned but almost unreadable; Solomon D. Goitein, *A Mediterranean Society*, 6 vols. (Berkeley: University of California Press, 1967–93), verbose but fascinating; Simon M. Dubnow, *A History of the Jews in Russia and Poland from the Earliest Times* (New York: KATV, 1975; originally published in Yiddish, 1920), a classic not yet superseded;

Moshe Idel, *Kabbalah: New Perspectives* (New Haven: Yale University Press, 1988), murky but important; Stephen Sharot, *Messianism, Mysticism, and Magic* (Chapel Hill: University of North Carolina Press, 1982), brilliant and incisive. No one can seriously involve himself in medieval Jewish history without reading a sample of books by two great Hebrew University scholars: Yitzhak F. Baer, *Galut* (Lantham, Md.: University Press of America, 1988; originally published in Hebrew, 1974); Gershom Scholem, *Kabbalah* (New York: Dorset, 1987, originally published in Hebrew, 1947). Among Gentiles who have written on Christian anti-Semitism, a substantial genre of its own, the best book remains Friedrich Heer, *God's First Love* (New York: Weybright and Talley, 1970). The volumes on the Middle Ages in Heinrich Graetz's classic eight-volume *History of the Jews*, originally published in German in the 1870s, of which there are many editions and translations (the Yiddish translation is much better than the English, not surprisingly), are still very much worth reading. There was nothing wrong with Graetz's medieval research: It is his ethnic chauvinism and implacable hostility toward the Catholic Church that is problematic.

SPECIALIZED STUDIES ON THE BLACK DEATH

Mark Ormrod and Phillip Lindley, eds., *The Black Death in England* (Stamford, U.K.: Paul Watkins, 1995). A general introduction by Jeremy Goldberg—inconclusive but suggestive and worthy of perusal—followed by Mark Ormrod, the learned historian of the reign of Edward III, on "Government in England after the Black Death." Anything Ormrod writes on fourteenth-century England merits careful consideration. But Ormrod's belief that the crown abandoned control over the legal system is unconvincing. Compare N. F. Cantor, *Imagining the Law* (New York: HarperCollins, 1997).

Robert C. Palmer, *English Law in the Age of the Black Death: A Transformation of Governance and Law* (Chapel Hill: University of North Carolina Press, 1993). Palmer assiduously reads court records, and what he has to say therefore merits close consideration. He argues just the opposite of what Ormrod claims. Palmer believes that the disorder attendant on the Black Death drove the royal government to *increase* central control of law and administration, using parliamentary statutes. The problem of the legal and political consequences of the Black Death may derive from the intrinsic nature of medieval public operations: Both ambitious efforts at central control and laxity and decentralization in practice happened at the same time.

Rosemary Horrox, ed., *The Black Death* (Manchester University Press, 1994), a collection of source material, well translated with extensive notes. There are too many narrative and rhetorical selections as compared to documentary sources.

BIOMEDICAL ASPECTS

William H. McNeill, *Plagues and Peoples* (New York: Anchor, 1998; originally published in 1976, new preface, 1998). The University of Chicago's intrepid and resourceful world historian stirred up great interest in biomedical history with this lively pioneering work that inspired a younger generation of historians to pursue the social history of pandemics. McNeill thought the Mongols, their migrations and conquests, were a key to plague history; there may be something in that.

Carol Rawcliffe, *Medicine and Society in Later Medieval England* (London: Sutton, 1997, first published in 1996). This is a sober and careful discussion, a first-rate piece of historical reconstruction. *Epidemics and Ideas*, T. Ranger and P. Slack, eds. (Cambridge University Press, 1992), contains some interesting perspectives and useful information.

Graham Twigg, *The Black Death: A Biological Reappraisal* (London: Batsford, 1984). This is the most important book, by a British zoologist, ever published on the biomedical history of the Black Death. I cannot see how anyone who has read this book carefully can persist in believing that the Black Death was exclusively bubonic plague. How much it was anthrax, as Twigg believed, is moot, but it is likely that anthrax or some similar murrain (cattle disease) was involved.

Fred Hoyle and Chandra Wickramasinghe, *Our Place in the Cosmos: The Unfinished Revolution* (London: Phoenix, 1996, originally published in 1993). Infectious diseases as well as human life came from outer space. Fred Hoyle of Cambridge is one of the world's leading astrophysicists, a great scientist.

Paul Davies, *The Fifth Miracle: The Search for the Origin and Meaning of Human Life* (New York: Simon & Schuster, 1999). Davies is a theoretical physicist and the recipient of the 1995 Templeton Prize for science writing; a serious writer, not a crank. He also asserts that infectious diseases and human life came from outer space.

Paul W. Ewald, *Evolution of Infectious Disease* (New York: Oxford University Press, 1994). The distinguished American biologist establishes conclusively the origin of major infectious diseases in Africa. This book is difficult but important.

OUT OF AFRICA

The great debate among paleontologists about where and when human life began appears to be over. Nearly all of these scientists now believe that human life, and subsequent threats to human life from infectious diseases, began in east Africa, near the borders of Kenya, Tanzania, and Ethiopia, around 2.5 million years ago. Among the many recent books on this subject, the most important are Virginia Morrell, *Ancestral Passions: The Leakey Family and the Quest for Mankind's Beginnings* (New York: Simon & Schuster, 1995), a neglected masterpiece and in a class by itself for readability; Roger Lewin, *The Origin of Modern Humans* (New York: Scientific America, 1993); Colin Tudge, *The Time Before History* (New York: Scribner, 1996), constantly thought-provoking and brilliantly written; Richard Leakey, *The Origin of Mankind* (New York: Basic Books, 1994); Donald Johnson and Blake Edgar, *From Lucy to Language* (New York: Simon and Schuster, 1996).

RECENT BIOMEDICAL PERSPECTIVES

Three recent books on current and recent biomedical research throw some intriguing light back on medieval infectious disease. These are Jeanne Guillemin, *Anthrax. Investigation of a Deadly Outbreak* (Berkeley: University of California Press, 1999); James Le Fanu, *The Rise and Fall of Modern Medicine* (London: Little, Brown, 1999); Gina Kolotka, *Flu* (New York: Farrar Strauss & Giroux, 1999).

Kolotka's book is a dramatic narrative of the still-unsuccessful effort to determine what the so-called Spanish influenza pandemic of 1918 was actually about. Guillemin's book indicates that the Soviet Union did not stop producing biochemical weapons in 1972 as the American government claimed the United States did. Soviet production of anthrax continued secretly at least until 1990 and may still be going on. Le Fanu's fascinating book is about the new medicine that began in the 1940s with penicillin and other antibiotics and the achievements and limitations of this new medicine. This is the best book on biomedical science and organization I have ever read. In *Betrayal of Trust: The Collapse of Global Public Health* (New York: Hyperion, 2000), the prize-winning biomedical journalist Laurie Garrett excoriates many recent governments for inadequate support of medical efforts to combat the growing number of pandemics. Ronald Reagan and his government come in for special condemnation. The book's approach is more narrative than analytical. It includes an account of a severe outbreak of pneumonic plague in India in the 1990s.

Stephen Porter, *The Great Plague* (Gloucestershire: Sutton, 1999), which is about the plague in London in 1665, has some relevance to the Black Death. The best part of the book is the pictures.

FILM

Aside from Ingmar Bergman's 1957 masterpiece, *The Seventh Seal*, there are two films that focus on the plague. The first half of the 1988 New Zealand film *The Navigator: A Medieval Odyssey* takes place in a mining village in Cumbria (northwest England) at the time of the Black Death. The impact of the plague and its terror is very well illustrated. The 1950 Hollywood film *Panic in the Streets*, starring Richard Widmark as a U.S. Public Health doctor, is about the outbreak of plague in New Orleans. The anxiety and confusion this causes has relevance to today and the future.

Acknowledgments

I wish to thank Dr. Anthony Gross for assisting me with valuable research in English and French documents. I also wish to thank my editors at The Free Press/Simon & Schuster, Bruce Nichols and Daniel Freedberg, and my literary agent, Alexander Hoyt, for encouragement and advice.

A decade ago, while I was Fulbright Professor at Tel Aviv University, I personally discussed some of the issues in this book with the masterful social historian of fourteenth-century England Professor Zvi Razi, whose studies of fourteenth-century English peasant life opened new horizons.

My secretary, Eloise Jacobs-Brunner, helped not only by preparing the disk for the publisher, but also with some valuable library research. I wish to thank the staff of the Bobst Library of New York University and the Firestone Library of Princeton University for their unfailing cooperation and courtesy. The dean of the College of Arts and Science at NYU helped me with research funding.

I am grateful to Edward I. Thompson of the University of Toronto for letting me read his unpublished paper "Another Look at Anthrax and the Black Death" and Gunnar Karlsson of the University of Iceland for allowing me to read his unpublished paper "Exterminating Rats in Medieval Plague Studies." Both papers were presented at a medieval studies conference at the University of Leeds in July 1999.

I wish to thank Michael Clanchy and Michael Prestwich for their valuable comments on earlier drafts of this book and their encouraging assessments of the project. William Beers and Brian Patrick McGuire also provided insightful critiques.

I wish to thank Nancy Silver Shalit for her very helpful comments on an earlier draft, and Dr. Henry Krystal, Emeritus Professor of Psychiatry, Michigan State University, for graciously allowing me to read two of his illuminating papers on memory of cataclysms.

Index

POCKET
B O O K S

SILENT NIGHT
The Remarkable Christmas Truce
of 1914
Stanley Weintraub

Silent Night brings to life one of the most unlikely
and touching events in the annals of war.

The First World War had been underway for only a
few months but had already been stalemated into
the brutality of trench warfare. As Christmas
approached, men on both sides – Germans, British,
Belgians and French – laid down their arms and
joined together in a spontaneous celebration with
their enemy. For a brief, blissful time a world war
stopped.

In *Silent Night*, Stanley Weinstraub brilliantly
conveys this strange episode, which is also one of
the most extraordinary of Christmas stories.
Tellingly, he also examines what might have
happened if the soldiers had been able to continue
the truce.

ISBN 0 684 86621 8
PRICE £12.99

POCKET
BOOKS

KINGDOM OF THE ARK

The Startling Story of How the Ancient British Race is Descended from the Pharaohs
Lorraine Evans

Kingdom of the Ark is the tale of a lost princess. A member of the Egyptian royal family and daughter of the 'heretic pharaoh', Akhenaten, and his wife, Nefertiti, Princess Scota fled her homeland, persecuted and friendless and landed on the shores of Britain. In this real-life historical detective story Lorraine Evans follows in the footsteps of these ancient mariners. By piecing together the archaeological clues and using genetic evidence, linguistic studies and Egyptology, she uncovers vital evidence to challenge the current view of accepted history and unceils the true origins of the ancient British race.

'Outrageously controversial but a genuine case has been made – no wonder there is a smile on the face of the Sphinx!' *Books Magazine*

ISBN 0 671 02956 8
PRICE £8.99

Norman F. Cantor is Emeritus Professor of History, Sociology and Comparative Literature at New York University. His academic honours include appointments as a Rhodes Scholar, Porter Ogden Jacobus Fellow at Princeton University, and Fulbright Professor at Tel Aviv University. Previous books include *Inventing the Middle Ages* and *The Civilization of the Middle Ages*, the most widely-read narrative of the Middle Ages in the English language. He lives in South Florida.